JOURNAL OF SCIENTIFIC EXPLORATION
A Publication of the Society for Scientific Exploration

(ISSN 0892-3310)

Editorial Office: *Journal of Scientific Exploration*, Society for Scientific Exploration, Kathleen E. Erickson, *JSE* Managing Editor, P. O. Box 1190, Tiburon, CA 94920 USA EricksonEditorial@att.net, 1-415-435-1604, (*fax* 1-415-435-1654)

Manuscript Submission: Submit manuscripts online at http://journalofscientificexploration.org/index.php/jse/login

Editor-in-Chief: Stephen E. Braude, University of Maryland Baltimore County, MD

Managing Editor: Kathleen E. Erickson, Tiburon, CA

Associate Editors
Carlos S. Alvarado, University of Virginia Health System, Charlottesville, VA
Daryl Bem, Ph.D., Cornell University, Ithaca, NY
Courtney Brown, Emory University, Alanta, GA
Prof. Etzel Cardeña, University of Lund, Sweden
Bernard Haisch, Digital Universe Foundation, USA
Michael Ibison, Institute for Advanced Studies, Austin, TX
John Ives, Samueli Institute, Alexandria, VA
Roger D. Nelson, Princeton University, Princeton, NJ
Dean I. Radin, Institute of Noetic Sciences, Petaluma, CA
Mark Rodeghier, Center for UFO Studies, Chicago, IL
Dr. Michael Sudduth, San Francisco State University, CA

Book Review Editor: P. D. Moncrief (pdmoncrief@yahoo.com)

Proofreader: Elissa Hoeger, Princeton, NJ

Society for Scientific Exploration Website — http://www.scientificexploration.org

Chair, Publications Committee: Robert G. Jahn, Princeton University, Princeton, NJ

Editorial Board
Dr. Mikel Aickin, University of Arizona, Tucson, AZ
Prof. Rémy Chauvin, Sorbonne, Paris, France
Prof. Olivier Costa de Beauregard, University of Paris, France
Dr. Steven J. Dick, U.S. Naval Observatory, Washington, DC
Dr. Peter Fenwick, Institute of Psychiatry, London, UK
Dr. Alan Gauld, University of Nottingham, UK
Prof. Richard C. Henry (Chairman), Johns Hopkins University, Baltimore, MD
Prof. Robert G. Jahn, Princeton University, NJ
Prof. W. H. Jefferys, University of Texas, Austin, TX
Dr. Wayne B. Jonas, Samueli Institute, Alexandria, VA
Dr. Michael Levin, Tufts University, Boston, MA
Dr. David C. Pieri, Jet Propulsion Laboratory, Pasadena, CA
Prof. Juan Roederer, University of Alaska–Fairbanks, AK
Prof. Kunitomo Sakurai, Kanagawa University, Japan
Prof. Yervant Terzian, Cornell University, Ithaca, NY
Prof. N. C. Wickramasinghe, Cardiff University, UK

SUBSCRIPTIONS & PREVIOUS JOURNAL ISSUES: Order forms on pages 383–384, or scientificexploration.org.

COPYRIGHT: Authors retain copyright to their writings. However, when an article has been submitted to the *Journal of Scientific Exploration* for consideration, the *Journal* holds first serial (periodical) publication rights. Additionally, the Society has the right to post the published article on the Internet and make it available via electronic and print subscription. The material must not appear anywhere else (including on an Internet website) until it has been published by the *Journal* (or rejected for publication). After publication in the *Journal*, authors may use the material as they wish but should make appropriate reference to prior publication in the *Journal*. For example: "Reprinted from [or "From"] [title of article]", *Journal of Scientific Exploration*, vol. [xx], no. [xx], pp. [xx], published by the Society for Scientific Exploration, http://www.scientificexploration.org."

Journal of Scientific Exploration (ISSN 0892-3310) is published quarterly in March, June, September, and December by the Society for Scientific Exploration, P. O. Box 1190, Tiburon, CA 94920 USA. Society Members receive a subscription to the *Journal* with their membership. Library and organizational annual subscriptions are $135. POSTMASTER: Send address changes to: *JSE*, P. O. Box 1190, Tiburon, CA 94920 USA

JOURNAL OF SCIENTIFIC EXPLORATION
A Publication of the Society for Scientific Exploration

Editorial

177 Editorial — STEPHEN E. BRAUDE

Research Article

181 Importance of a Psychosocial Approach for a Comprehensive Understanding of Mediumship — EVERTON MARALDI, FATIMA REGINA MACHADO, WELLINGTON ZANGARI

Essays

197 Investigating Mental Mediums: Research Suggestions from the Historical Literature — CARLOS S. ALVARADO

225 Advantages of Being Multiplex — MICHAEL GROSSO

247 Some Directions for Mediumship Research — EMILY WILLIAMS KELLY

Historical Perspective

283 Parapsychology in France after May 1968: A History of GERP — RENAUD EVRARD

Letters to the Editor

295 Response to "How To Improve the Study and Documentation of Cases of the Reincarnation Type? A Reappraisal of the Case of Kemal Atasoy" — H. H. JÜRGEN KEIL, JIM B. TUCKER

297 Reply to Tucker and Keil — VITOR MOURA VISONI

297 Weight of the Soul: 28 Grams? — NILS O. JACOBSON

298 Human Weight Loss upon Death — FRANK G. POLLARD

Obituary

299 Rémy Chauvin (1913–2009) — RENAUD EVRARD

Book Reviews

305 *The Tunguska Mystery* by Vladimir Rubtsov — STEVEN J. DICK

307 *Witness to Roswell: Unmasking the Government's Biggest Cover-Up* by Thomas J. Carey & Donald R. Schmitt — BRUCE MACCABEE

310 *The Roswell Legacy: The Untold Story of the First Military Officer at the 1947 Crash Site* by Jesse Marcel, Jr., & Linda Marcel — BARRY GREENWOOD

315 *Art, Life, and UFOs: A Memoir* by Budd Hopkins — STUART APPELLE

319 *Bigfoot: The Life and Times of a Legend* by Joshua B. Buhs; *Anatomy of a Beast: Obsession and Myth on the Trail of Bigfoot* by Michael McLeod — JOHN BINDERNAGEL

323 *Philosophy in the Flesh: The Embodied Mind and Its Challenges to Western Thought* by George Lakoff and Mark Johnson — STEPHEN R. PALMQUIST

327 *Neither Brain nor Ghost: A Non-Dualist Alternative to the Mind–Brain Identity Theory* by W. Teed Rockwell; *Radical Embodied Cognitive Science* by Anthony Chemero — STAN V. MCDANIEL

336 *Synchronocity: Multiple Perspectives on Meaningful Coincidence* edited by Lance Storm — FRANK PASCIUTI

341 *Laboratories of Faith: Mesmerism, Spiritism, and Occultism in Modern France* by John Warne Monroe — RENAUD EVRARD

345 *The Spiritual Anatomy of Emotion: How Feelings Link the Brain, the Body, and the Sixth Sense* by Michael A. Jawer with Marc S. Micozzi — BRYAN J. WILLIAMS

351 *La Connaissance Supranormale, Étude Expérimentale* by Eugène Osty. *Supernormal Faculties in Man: An Experimental Study* by Eugene Osty — THOMAS RABEYRON

358 *Phénomènes Psychiques au Moment de la Mort* by Ernest Bozzano; *Deathbed Visions: The Psychical Experiences of the Dying* by Sir William F. Barrett — CARLOS S. ALVARADO

364 *Allan Kardec und der Spiritismus in Lyon um 1900. Geisterkommunikation als Soziales Phänomen* by Katrin Heuser — ANDREAS SOMMER

366 Further Book of Note: *El Mundo Oculto de Los Sueños* by Alejandro Parra — CARLOS ADRIÁN HERNÁNDEZ TAVARES, STANLEY KRIPPNER

SSE News

367 Announcement for 2010 grants from the Helene Reeder Memorial Fund for Research into Life after Death; SSE 30th Annual Meeting, 2011 in Boulder

368 SSE Masthead

369 Index of Previous Articles in *JSE*

381 Order forms for *JSE* Issues, *JSE* Subscriptions, and Society Membership, Instructions for Authors

EDITORIAL

This issue of the Journal is devoted to a single multifaceted topic: mediumship, and mental mediumship in particular. As many readers will know, mental mediumship is usually contrasted with physical mediumship, and the distinction—roughly—is this. In the former, the medium channels communications from other, presumably nonphysical, entities, whereas in the latter the medium channels phenomena that some would classify as ostensibly psychokinetic (e.g., object movements and levitations, or materializations). In both cases, however, the received view among practitioners and other believers is that the medium is an intermediary between our familiar physical world and entities occupying a different order of reality. In the former case, mediums purport to facilitate spirit *communication*, and in the latter case, spirit *agency*.

Like many topics usually classified as parapsychological, mediumship can bring out the best and the worst in scientists and other scholars. Some reflexively and uncritically dismiss a belief in mediumship (or even mere openness to belief) as deplorably atavistic, embracing or entertaining a crude and primitive superstition clearly unworthy of serious attention. In fact, some would quickly denounce as intellectually defective anyone taking a serious scholarly interest in the subject. Now of course it's true that people have held mediumistic beliefs for millennia and in quite different cultural contexts, many of them widely considered to be primitive or unsophisticated. Nevertheless (and equally clearly), we're not entitled to reject those beliefs simply in virtue of their longevity and pervasiveness. Morever, the same critics who scoff instinctively at mediumship often fail to recognize that their aversion to the topic may be equally reflexive and uncritical. In fact, it may betray its own suspicious and venerable lineage, which F. C. S. Schiller once described as "the instinctive dislike which everywhere has led to the prohibition of 'sorcery', to the burning of 'witches', and to the ascription of the phenomena generally to the agency of the devil" (Schiller, 1899:105).

By contrast, other writers see mediumistic phenomena as a potential source of insights, not into other realms, but into many aspects of antemortem human nature. Of course (as I noted), mediumship superficially presents itself as a form of interaction between our familiar physical world and another world of spirits, invisible to all but a privileged (or afflicted) few. That's certainly the prevailing view among practitioners and their followers, and it's also the view that parapsychologists have scrutinized painstakingly ever since the founding of the Society for Psychical Research (SPR) in 1882. Indeed, to this day a major preoccupation of the SPR is to document mediumistic claims carefully and critically, determine whether (or to what extent) those claims are veridical, and then consider whether the veridical claims deserve to be taken as evidence for postmortem survival.

But it's not only spiritists, spiritualists, or psychical researchers who take mediumship seriously. It's also a rich field of research to intrepid (or professionally secure) anthropologists, psychologists, psychiatrists, historians, physiologists, and even the occasional philosopher (I was going to say "odd philosopher," but I realize that many would regard that expression as redundant). Consider this. Some researchers contend that mental mediumship looks very much like a culturally variable expression of the same underlying and unusual human capacities studied in many other contexts, both pathological and nonpathological. For example, they're quick to note similarities between mediumship and forms of nonpathological dissociation or hypnosis, and (even more dramatically) ostensible demonic possession, dissociative identity (or multiple personality) disorder, and psychogenic fugue, which likewise assume appropriately different forms in different epochs and cultures. If these researchers are right, the forms of mental mediumship might plausibly be viewed as symptom-languages or idioms of distress, but nevertheless as ways of unleashing or accessing otherwise latent and perhaps exceptional human capacities. And if that's right, mediumship might have much to teach us about the nature, variety, and limits of human abilities generally. And that's just one nonparapsychological way to approach the topic.

The papers in this issue of the *JSE* examine mental mediumshp from several angles. Maraldi et al. describe several varieties of mediumship, and propose that the psychosocial and cultural aspects of mediumship are as important as its psychophysiological features. They consider the differences between spiritism and spiritualism, the relevance of the trance state to mediumship, and the extent to which processes properly considered mediumistic infiltrate many everyday human activities. By drawing in particular on some recent case studies, the authors focus on the spiritistic views of Allan Kardec, according to which mediumship is a pervasive and fundamental feature of human nature and whose study is therefore essential to the behavioral sciences.

Carlos Alvarado helpfully surveys some of the major nonparapsychological issues that have fascinated researchers in mediumship. These include the nature and prevalence of the mediumistic trance, the varieties of mediumistic mentation, the evolution of mediumship over time, the dramatic capabilities of the subconscious, and the relationship between mediumship and psychopathology.

Michael Grosso focuses on positive aspects of mediumship, and in particular its connection with various forms of artistic and practical creativity. For example, he describes some dramatic cases of mediumistic writing and painting, and with reference to F. W. H. Myers's theory of genius he considers mediumship's potential for personal development or transformation.

Emily Kelly's paper considers the issue that motivated the pioneers of the SPR: whether mediumship ever indicates that human personality survives bodily death. She recognizes that the debate over that key issue hangs on whether the best cases should be interpreted not as evidence for postmortem survival but as evidence for psychic functioning among the living. (For some recent discussion in this *Journal* on that subject, see Sudduth, 2009, and Braude, 2009.) Kelly surveys types of mediumship that have been especially relevant to that debate, and advocates the revival of so-called proxy sittings with mediums. In this type of experiment, the person sitting with the medium is a proxy or stand-in for the person actually desiring information from the medium. And in the best of those cases, the proxy is not personally acquainted with the distant "sitter."

Finally, Renaud Evrard's paper doesn't discuss mediumship directly. Rather, it recounts a period in France's recent history during which a group of researchers rejected the laboratory-based experimental approach advocated by Rhine and others, lobbying instead to have parapsychology broadened into a multidisciplinary field that embraces (in additional to strictly experimental and proof-oriented investigations) wide-ranging enquiries that could be applied to mediumship and other apparent instances of psi-in-life.

These papers by no means exhaust the fascinating and important topics connected with the study of mediumship, although they comfortably fill this issue of the *JSE*. Since we have other excellent papers on mediumship currently in our editorial pipeline, I look forward to publishing another special issue on mediumship in the near future.

STEPHEN E. BRAUDE

References

Braude, S. E. (2009). Perspectival awareness and postmortem survival. *Journal of Scientific Exploration 23*(2), 195–210.
Schiller, F. C. S. (1899). Review of F. Podmore's *Studies in Psychical Research*. *Mind 8*, 101–108.
Sudduth, M. (2009). Super-psi and the survivalist interpretation of mediumship. *Journal of Scientific Exploration 23*(2), 167–193.

Importance of a Psychosocial Approach for a Comprehensive Understanding of Mediumship

EVERTON MARALDI

FATIMA REGINA MACHADO

Laboratory of Research for Social Psychology of Religion

WELLINGTON ZANGARI

Laboratory of Research for Social Psychology of Religion
w.z@usp.br

All authors: InterPsi—Laboratory for Anomalistic Psychology and Psychosocial Processes,
University of Sao Paulo, Department of Social Psychology,
Av. Prof. Mello Moraes, 1721, Sao Paulo SP 05508-900, Brazil

Abstract—There are several definitions of mediumship. The majority of them are religion-based. In this article, the term *mediumship* is defined as the supposed capacity that certain people—that is mediums—are said to have by which they can mediate communication between spiritual entities or forms and other human beings. Such a definition does not explain the origin of mediumship, but rather highlights its characteristics as they are reported by people who experience the phenomena in different sociocultural contexts. In general, it is said that mediumistic capacity is aroused when the medium is in an altered state of consciousness such as a trance state. However, for Kardecist Spiritists for example, mediumship may also occur in conscious states. Mediumship can be present in practically any human activities, from the elaboration of a scientific or literate text to an artistic production, as well as in such minor experiences as vague physical sensations or even emotional states such as irritability, sadness, sudden joy, obsessive thoughts, moments of inspiration or geniality, and so on. In all these experiences—from the most common to the most exceptional—Kardecist Spiritists admit the possibility of spirit intervention. So, in many cases there is no clear delimitation between what comes from the medium as an individual and what would come from an external source. Although the Kardecist perspective—very widespread in Brazil—is based upon certain religious and philosophic hypotheses that are unacceptable for many scientists and academics, Kardecism has contributed to the development of scientific and psychological conceptions of so-called mediumistic manifestations. Shamdasani (1994:xiv) pointed out that because Kardec believed that mediumship was a fundamental aspect of humanity and must be considered in order to understand the human condition, his Spiritist doctrine was formed in such a way as to facilitate psychological interpretation of what seemed to be mediumistic phenomena. The main difference between Kardec's theory and a completely psychological study of mediumship is the cause of the phenomena: that is whether the phenomena occurred through the action of spirits or through the action of the medium's subliminal or subconscious imagination. The search for an intrapsychic source for mediumship contributed to the "discovery" of the unconscious mind.

Keywords: mediumship—Spiritism—psychosocial identity

Approaches to the Study of Mediumship, Their Contributions and Consequences

Proper scientific studies of mediumship were first conducted in the period between the late 19th century and the early 20th century, arousing huge public interest, especially in the United States and Europe. However, the interest in the study of mediumship had already appeared in the 18th century when Christian values were profoundly affected by the emergent worldview that was based on industrial and scientific development. These new ideas threatened the ontological status of beliefs in the existence of life after death (Northcote, 2007). Rationalism and the concomitant growth of positivism became popular, imposing a strictly materialistic conception of reality. The new way of looking at the world conflicted with previous religious conceptions, especially those that touched on the origin of human beings and the existence of the soul (Ronan, 2001). Society began to move away from a religious worldview based simply on religious dogmas. To some it seemed necessary to adopt scientific methods in order to scientifically prove the existence of the soul and its immortality (Rogo, 1986). But even if the majority of the scientific and academic community adopted a materialist view of the world and society as a whole questioned the traditional basis of religions, the masses began to look for a worldview that was both religious and, paradoxically, empirical (Northcote, 2007).

Scientific psychology appeared in that context. In the beginning, the discipline was strongly connected to the study of alleged paranormal experiences and especially to the study of mediumistic experiences (Alvarado 2005, Alvarado, Machado, Zingrone, & Zangari, 2007, Ellenberger, 1976). By the end of the 19th century, mediumship was the object of many psychological studies (Shamdasani, 1994). However, the majority of researchers saw mediumistic practices only as the fruit of frauds or as a dangerous threat to the well-being of society. Mediumship was commonly linked to psychopathology in the psychiatric literature of the times (Le Maléfan, 1999). There were physicians who maintained that brain damage and/or other functional disturbances caused spiritualist/Spiritist beliefs and practices or vice versa. Many of those physicians believed that spiritualist beliefs were a principal cause of insanity (Hess, 1991, Shamdasani, 1994, Zingrone, 1994). Consequently, people who practiced Spiritism or spiritualism or who participated in seances began to be persecuted in Europe and in such countries as the United States and Brazil (Almeida et al., 2005, Giumbeli, 1997, 2003, Machado, 1996, 2005). With the influx of Social Darwinism, the psychiatric community began to use the metaphors of evolution, characterizing mediumistic manifestations along the scale of intellectual and social development.

According to Zingrone (1994), behind the intrigues involving mediums and scientists, there were racial, gender, and social status conflicts, as well as conflicts stemming from political and religious interests. Mediumistic phenomena were associated with such marginal social groups as women, individuals of African descent, and the poor, reproducing and amplifying existing prejudices. Some criticisms aimed at Spiritists/spiritualists had an evident party–political content. They alluded in a pejorative way to the connection of mediumship to suffragism and other social movements. Critics also were against beliefs incompatible with Roman Catholic dogma—which was considered a model of religious institutional beliefs, especially in the West—and against several forms of alternate therapies not approved by the medical establishment. Efforts to stop spiritualists and Spiritists could take a judicial form. As Zingrone says:

> For the anti-Spiritualists, this characterization of Spiritualism and mediumship as categories of mental illness was deadly serious and could be used to strip "the patient" of legal rights and social privileges. Owen (1990) uncovered an abundance of legal records of "lunacy" cases tried in England. A number of similar cases were litigated in the United States as well [Haber, 1986]. The evidence in some of these cases rested solely on defendants' interest in, practice of, or belief in Spiritualistic phenomena. Purely on this basis, many persons, mostly women, were involuntarily committed to asylums. Only a handful of these inmates were able to obtain their release, and then only after enlisting the legal and financial aid of other more socially powerful Spiritualists (Owen, 1990:160–167, 168–201). Many others—again, mostly women—lost legal control of their monetary resources to members of their immediate families—usually men—on the grounds that commitment to Spiritualism was symptomatic of a severe chronic mental incapacity (Owen, 1990:160–167). (Zingrone, 1994:102–103)

In spite of the persecutions, however, a segment of the scientific community considered mediumship from another point of view. In 1882, a group of intellectuals from different areas decided to join forces to found *The Society for Psychical Research* (SPR), the purpose of which was to investigate so-called mediumistic and paranormal events using the scientific method.[2] Investigation on mediumship done by the SPR resulted in pioneering studies on dissociation and altered states of consciousness (Alvarado, 2002).

The first studies of mediumship were strongly influenced by their historical time. People then were still deeply involved in their traditional religious beliefs, so much so that many scientists still believed in life after death, and felt it was possible to study the topic from a scientific point of view. Similarly, such scientists were also interested in studying mediums. Frederic Myers, for instance, was both favorable to the Spiritist survival hypothesis and explored evidence related to it in his work (Myers, 1903/2001:31). Theodore Flournoy, who was skeptical of the possibility of communication with the dead, declared that, even though he felt that way, he still considered himself a Spiritualist and therefore believed in the existence of a transcendent dimension in the human being. As Flournoy said:

> Let me insist here that we must not confound *Spiritism*, which is a pretended scientific explanation of certain *facts* by the intervention of spirits of the dead, with *Spiritualism*, which is a religio–philosophical belief, opposed to materialism and based on the principle of value and the reality of individual consciousness, and which I conceive to be a necessary postulate for a wholesome conception of the moral life. [...] One may be a Spiritist without being a Spiritualist, and *vice versa*. So far as I myself am concerned I am a convinced Spiritualist. (Flournoy, 1911/2007:142)

Although pioneers of scientific studies on mediumship had different opinions and beliefs on the subject, all of them privileged the scientific posture in their studies and shared some ideas: (a) they believed in the importance of the psychological study of mediumship for its potential to add to our understanding of the human mind; (b) they understood that mediumship is a complex phenomenon but despite that fact it was possible to formulate certain testable hypotheses for mediumistic phenomena that were not sufficiently well-understood; (c) they believed that even though mediumship could sometimes be associated with psychopathology, they felt that such an explanation could not be applied in general because it did not explain all the available evidence; and (d) while the majority of them were in doubt about the paranormal nature of some of the phenomena described in the mediumistic context, in other cases available evidence confirmed the legitimacy of the paranormal hypothesis.

The majority of the early investigations conducted by SPR members were focused primarily on verifying that the content of supposed mediumistic communications had, in fact, originated with the deceased. Although many hypotheses proposed by these researchers contained psychological concepts, their main purpose was to investigate the notion of survival (Zangari, 2003). One evident problem in these early investigations was the emphasis on individual aspects of mediumship, that is on the study of intrapsychic and unconscious processes in mediums. This approach neglected the power of culture and society in modeling beliefs and experiences associated with mediumistic phenomena (Maraldi, 2008).

In the first two decades of the 20th century, the interest in mediumship diminished considerably. The ascension of psychoanalysis has been said to have been greatly responsible for the decline. Unlike psychical research—an empirical discipline—psychoanalysis established therapeutic methods that became popular and, to some extent, obscured the investigation of mediumistic phenomena and other forms of paranormal experiences as verifiable occurrences. Psychoanalytic techniques made possible a more controlled and rational relationship with the unconscious, less intense and passionate than those observed in mediumistic seances (Alvarado, 2002, Shamdasani, 1994).

Due to the association of mediumship with Spiritism/Spiritualism and Occultism, mediumistic phenomena began to be considered threatening to psychology because their studies were seen as metaphysical speculations. Psychology gradually abandoned its interest in mediumship, concentrating instead mainly on psychopathology, learning, child development, and comparative psychology, and other areas of modern psychology. Theories began to focus on referential models designed to increase comprehension of the psyche. Mediumship was set aside and only recently has become the object of intensive investigations on a par with those conducted previously (Almeida & Lotufo Neto, 2004:137).

The renewed academic interest in mediumship is relatively recent. It has been occurring because of the growing enthusiasm about the study of paranormal beliefs and/or experiences in general. Mediumship studies have made little progress over the intervening decades, and researchers continue to face practically the same problems faced by the pioneers of mediumship research.

Perhaps the most important contribution of recent work is the establishment of mediumship as a psychosocial phenomenon, which, in a certain way, disconnects it from exclusively intrapsychic and psychopathological interpretations. Regarding the psychosocial aspect, modern researchers have understood that perceiving mediumship as a dissociative phenomenon constrained by the historical and social context was something that most of the pioneer researchers missed because of their focus on individual aspects of the mediums themselves. As Zangari has said:

> [...] in spite of the fact that mediumship "uses" a medium's dissociative capacities, it seems to be a dissociation disciplined by the medium's social group. [...] Socio-cultural elements that outline the "intruder" personality are present in the medium's social group and, therefore, in the medium's mind [. . .] the difference between pathological dissociation and non-pathological dissociation lies in the consideration of context, of culture. (Zangari, 2003:54–55).

Unilateral analyses of mediumship tend to partial—and therefore incomplete—interpretations. Monique Augras (1983:77) has criticized those who have tried to understand mediumship using an approach that excludes the cultural elements involved in the phenomenon.

There are clinical studies that have considered mediumship from psychosocial perspectives that seem to corroborate the importance of mediumship as a social construction as opposed to seeing the phenomena from a merely pathological or intrapsychic point of view. Several authors have provided evidence for the notion that mediumship is not necessarily associated with psychopathologies (Almeida, 2004, Almeida, Lotufo Neto, &

Greyson, 2007, Negro, 1999, Reinsel, 2003). Grosso (1997) defends the notion that mediumship, artistic inspiration, and surrealism are forms of "creative dissociation". According to Grosso (1997) and Zangari (2003), what seem to be fragmentations and disintegration in a specific culture may be the prelude to a major psychological integration in another culture.

Discussing the correlation between mediumship and dissociative identity disorder (formerly known as multiple personality disorder), Braude (1988) has suggested that while the creation of multiple personality usually begins as a reaction to unbearable traumatic events, mediumship tends to develop in a more healthy way—although the author considered the possible link between certain mediumistic phenomena and psychopathology. A similar opinion was adopted by Richeport (1992).

After interviewing and administering dissociation scales to individuals diagnosed as suffering from multiple personality disorder, Hughes (1992:191) has concluded that mediums do not present a high level of psychopathology or dissociative experiences. Mediums differ from those who have multiple personality disorder in terms of mental processes (etiology, function, control, and psychopathology). For those who have multiple personality disorder, dissociation is compulsive; for mediums, it is disciplined and culturally contextualized.

Similarly, recent research results have not confirmed that mediumship is invariably a defense mechanism against psychological suffering or social exclusion (Negro, 1999). Almeida (2004), for instance, has demonstrated that the socio–demographic profile of Spiritists in Brazil is quite different from what is expected.

The Flourishing of a Psychosocial Perspective of Mediumship in Brazil

It is said that Brazil—often characterized as the largest Spiritist country in the world—is a *warehouse* of paranormal experiences, especially Spiritist-like ones. And Brazil is also well-known for its mixture of cultures and its tolerance of different religious beliefs. But it was not always like that. Spiritist practices were illegal at one point in the history of the country—especially in the first part of the 20th century—and during that time a number of interpretations of mediumship related to dissociation were proposed. But the cultural context of experiences was usually neglected. Mediumship was also almost invariably described as a symptom of psychopathology. Raimundo Nina Rodrigues (1862–1906), an eminent Brazilian physician and anthropologist, interpreted possession as "a provoked sleepwalking-like state with fragmentation and substitution of personality" (Nina Rodrigues, 1900:81). Manoel Querino (1851–1923), a Brazilian intellectual and anthropologist who was a pioneer historian of African culture in Bahia, Brazil, described mediumship as an impressive hysterical phenomenon peculiar to women (Querino, 1955:73).

Monique Augras (1983) points out that such important researchers and academics as Artur Ramos (1903–1949), a Brazilian physician, anthropologist, and folklorist, believed "that trance does not reveal any characteristic beyond what is already established by Psychiatry as mass hysteria" (p. 36). Brazilian researchers were influenced by 19th century European psychiatric ideas, especially those in France, such as Le Bon's concept of "mass hysteria" and Charcot's ideas on hysterical dissociation as a neurological degeneration of women (Ellenberger, 1976).

The opinion of the Brazilian medical community on *Spiritism*—a term that was commonly used to refer to Kardecist Spiritism as well as mediumistic African or Afro–Brazilian practices—was linked to the Brazilian historical context. Spiritists in general were persecuted for political and religious views when Getúlio Vargas was the President of Brazil (1930–1945 and 1951–1954). A large number of Spiritist centers were closed during those years (Hess, 1991). The campaign against Spiritism helped to legitimize the position of physicians who were against Spiritist beliefs and practices. For example, in the 1920s and 1930s, the *Liga de Higiene Mental* (League for Mental Hygiene) considered Kardecist Spiritism and other mediumistic religions to be both a mental health problem and a social problem (Costa, 1976, Oliveira, 1931, Ribeiro & Campos, 1931). Their views were consonant with the opinion of such early Brazilian psychoanalysts as Artur Ramos.

Social dimensions of mediumship started to be considered formally in Brazil with the work of Melville J. Herskovits (1967) who did not see mediumship only from a psychopathological point of view. According to Augras, Herskovits "states that ritual trance, while institutional, is an abnormal phenomenon. It is an organized cult, not an individual pathology" (Augras, 1983:47). Herskovits' perspective was assumed later by such Brazilian researchers as Octavio da Costa Eduardo and René Ribeiro (1978). However, it was the French sociologist Roger Bastide (1898–1974) and the French photographer and ethnologist Pierre Verger (1902–1996) who first proposed a properly sociological and historical perspective for the analysis of mediumistic religions in Brazil in the mid-1950s (Bastide, 1989, Verger, 2002).

Sociologists and anthropologists, more than members of any other discipline, have offered psychological or social interpretations for mediumistic phenomena. However, the sociological perspective is a generalist one and does not always adequately consider the individual and group dimensions in their interaction with broader social and institutional processes. Social psychology, on the other hand, takes on as its task the investigation of particularities and mechanisms involved in the dialectic between group and individual. From a social psychological perspective, then, mediumship can be seen as a psychosocial construction, that is as a phenomena that is simultaneously individual and collective. Recently, this perspective has been flourishing in Brazil and seems to us to be a fruitful approach to the study of mediumship.

An Exploratory Psychosocial Study of Mediumship in Brazil

An exploratory study conducted by the Brazilian psychologist Everton de Oliveira Maraldi (2008) is offered here as the first step in a larger research project focused on psychosocial aspects of mediumship in Brazil. His study aimed to understand the use and meaning of mediumship and the paranormal beliefs connected to it in the formation of the psychosocial identity of Kardecist Spiritists. The research program draws on the theory of identity proposed by Ciampa (1987, 1994), a well-known social psychologist in Brazil. Ciampa's work can be defined as a tentative recasting of identity theory that is anchored especially in Habermas's (1990) philosophy.

According to Ciampa, identity is in a constant state of transformation and metamorphosis, passing through different moral or cognitive stages of development. He also recognizes identity as a predominantly social phenomenon, that is all individuals contribute to the actualization of a group's identity even if it is only in a potential way. Individual particularities reproduce universal particularities. Thus, group identity and individual identity are not disconnected. They form together in a shared context. Identity can be understood as two different aspects in a dialectical relation: the representational aspect (categorization) and the dynamic aspect (metamorphosis), both seen as process. Identity results primarily from the process of metamorphosis itself, and its construction occurs across the lifespan. The construction of a "good" identity project is a process that can be favorably influenced (or not) by social conditions or by adaptive crises faced by the individual in his or her daily life.

Using Ciampa's approach to discuss the construction of Kardecist Spiritist mediums, Maraldi's exploratory research was composed of two "life history" case studies with two female mediums from the *Centro Espírita Ismael* (http://www.ceismael.com.br) in the city of São Paulo in Brazil. In addition to doing deep interviews with the mediums, Maraldi visited the Spiritist center several times and observed mediums in action. He has also considered other complementary materials for his analysis, such as psychographic messages and a mediumistic drawing. Maraldi has selected two mediums who seemed to have effectively established a significant personal and group connection to mediumistic beliefs and practices. The main objective of the exploratory study was to evaluate the potential influence of indoctrination and group context on the maintenance of beliefs and on the construction of mediumship as an identity, as well as to raise hypotheses for future investigations.

From the analysis of interviews—as well as consideration of the complementary material—using Ciampa's ideas as reference, Maraldi verified that the uses and meanings of mediumship in the construction of the investigated mediums' identities varied not only in terms of psychodynamic functions but also in psychosocial meanings. These can be reduced to the following three basic categories:

(a) ***Mediumship as a life project.*** This is a category that includes the *re-signification* of mediumistic and Spiritist identity in the interviewed mediums' life histories as the search for emotional and spiritual meaning. In this category, Maraldi explored how daily situations and physical and emotional experiences are interpreted by each of interviewees (E.D.E. and I.N.) to conform to their evolving mediumistic identities. That is, how the interviewees learned to give a spiritual meaning to their personal problems, difficulties, or even successes. Maraldi also investigated the interviewees' interpretations of their mediumistic abilities and their efforts to bring their experiences into conformance with Kardecian Spiritist doctrine as their mediumistic identity developed and matured, and as they came to see their mediumship, as they told Maraldi, as predestined and predetermined even from before their reincarnation into their present lives.

(b) ***Mediumship as a way to veil or unveil identity.*** This is the category in which Maraldi examined the psychodynamic means by which the interviewees masked or disclosed their mediumistic identity in the context of the Spiritist center including those moments when their activities occurred within unconscious mediumistic states, or through what seemed to be paranormal phenomena.

(c) ***Mediumship as ideology.*** In this category, Maraldi's investigation revolved around the notion that mediumship can be seen as an ideological posture that is a kind of materialization of Spiritist doctrine. He focused his analysis on the way in which the interviewees commonly incorporated scientific and religious tropes in support of Spiritism in their speech. In this category, Maraldi noted that the discourse in which the interviewees engaged repeatedly centered on the debates between materialist science and Spiritism, medicine and Spiritism, and Catholicism and Spiritism.

When dealing with "mediumship as a life project", Maraldi's results supported the notion that the formation of a mediumistic identity organized the mediums' emotional experiences so as to sustain a life project that was previously nonexistent or inconceivable. This is the re-signification function of mediumship: a search for meaning to face, explain, or justify difficulties or successes. Not only are certain psychodynamic functions serviced—such as diminishing anguish and anxiety from the exposure to conflictful and traumatic situations—but also future personal psychological needs are prepared for. From the analyses of the mediums' discourse, it becomes apparent that prior to their conversion to Spiritism and their training as mediums, the interviewees' lives had no defined direction, as they were unsure as to what role they might play in the world. The interviewees tended to see their pre-mediumship past as marked by problems that could only be solved after their predestined dedication to Spiritism. The way in which the mediums described their lives before mediumship centered on a painful life full of disease and disturbance. Both mediums had endured difficult childhoods. E.D.E. was ill frequently and hospitalized repeatedly without her doctors being able to render a diagnosis. I.N. had psychological and family problems. Her childhood was complicated by continually arguing parents to the extent that her siblings seemed to have played a more significant role in her upbringing than her parents did. Both E.D.E. and I.N. also suffered from relatives who considered them to be mentally disordered. On the other hand, the "discovery" of their mediumship re-signified their experiences, giving them a paranormal meaning. For E.D.E., what she thought were hallucinations became the perception of spirits. I.N.'s mood swings and other emotional difficulties were not only the result of family conflicts, but also flowed from her as-yet-untrained ability to "capture energies" from discarnate spirits or living people, something mediumship training helped her to balance. Seemingly inexplicable relationships and occurrences in both mediums' family lives were now understood as consequences of past lives. For instance, E.D.E. explained her difficult relationships with her parents and her good relationships with her grandparents as a result of her parents having been her aunt and uncle in a previous life while her grandparents had been her parents. An emotional dependency on her displayed by her brother was also explained in terms of past life events. In effect, E.D.E.'s attributions of past life causes to current life events or relationships allowed her to delineate a mythical origin to those elements—including social roles—of her current life that she considers *emotionally* unacceptable or incomprehensible.

Not only do mediumship practices help E.D.E. and I.N. to deal with their experiences, but social perceptions both made by and of the interviewees have also changed. As mediums, both E.D.E. and I.N. find themselves embedded in a context in which who they are and what they do are culturally valued. For I.N., for example, the "discovery" of her mediumship has been a transformative experience that has changed her from a shy and unstable person to someone who has occupied a more adaptive and fulfilling social role in life.

The second analytical stance Maraldi took with his interviewees was to follow Ciampa's (1987) notion that identity is alternatively veiled and unveiled. That is, an individual's multiple social roles are revealed or hidden at different moments, each role being only a partial representation of the whole individual. These transitions may be conscious or unconscious and it is the unconscious roles that may become strange to the waking self, dissociated because they cannot be openly assumed. A tension is created, however, in that what is masked and unconscious must seek to be disclosed, expressed in some way. In this sense, the context of *Centro Espírita Ismael* provides E.D.E. and I.N. with the opportunity to unveil an otherwise veiled aspect of their selves through exercising their mediumistic identity. The Spiritist center becomes a safe, controlled environment in which those aspects of the mediums' psychology that are repressed or undeveloped may be expressed (and we are not talking only about mediumistic ability surfacing: Maraldi also observed such creative processes as painting and writing being expressed in the context of the *Centro*). Diffuse emotions may also be expressed, and by doing so the mediums are able to deal with their subjective world without assuming total responsibility for what emerges in mediumistic sessions.

Mediums report they are susceptible to feelings that damage family and social life. To maintain an emotional balance she believes is necessary for her mediumship, I.N., for example, says she avoids feeling anger or hostility toward her husband. Allowing her emotional balance to be upset I.N. fears makes her vulnerable to invasion by something strange and compulsive that could harm her, her family, and her mediumship. She is also afraid of losing effective contact with the world, of dissociating from herself. I.N. fears possession as a sign of obsession, an uncontrollable negative influence by spirits. From a psychodynamic point of view, however, that which I.N. sees as an abrupt intrusion of the spirit world may be, in fact, an unconscious activity, what Jung (1920/2004) defined as "ideo-affective complexes". As Jung has said, complexes tend to become true secondary personalities, or, in other words, *unconscious roles*, such as the roles that mediums can freely experience in the Spiritist center, but for which they do not assume complete authorship.

In their own way, activities at the Spiritist center can be integrative and therapeutic, facilitating safe contact with the unconscious and promoting the adaptive development of individual identity through a form of emotional control instilled through doctrinal compliance. The Spiritist center provides meaning through both a symbolic system and practical training that allows the medium to interpret and control otherwise disturbing experiences without fear. The power of Spiritist doctrine to shape the medium's emotional control can be seen as a progressive capacity that increases in efficacy as the mediums develop. Zangari (2003) has called this process "the training of altered states of consciousness in a ritual context". Negro (1999), on the other hand, characterizes this process as the "modeling of social behavior" or the assimilation of experiences in a "social matrix".

Techniques employed at the *Centro Espirita Ismael* to train or induce mediumship are derivations of hypnosis that clearly evoke altered states of consciousness in which it is suggested that the emotions should be worked on through practices that comply with Spiritist doctrine. The capacity of mediums to *surrender* to the spirits—as proposed by Zangari (2003)—is a passive surrender to (potential) unconscious elements. The Spiritist center both welcomes and allows latent content within a socially accepted ritual that, at the same time, helps mediums preserve the stability and integrity of their personal identity.

However, mediumship practices also involve risks. An extreme adherence to Spiritist doctrine can provoke, to some extent, a resistance to change, to identity metamorphosis, resulting in the mere re-positioning of social roles and the exercise of repressive "control" of aspects deemed undesirable by, or incompatible with, Spiritist doctrine. During his visits to the Spiritist center, Maraldi noticed that it was not uncommon for participants to repress contrary phenomena not in keeping with Spiritist ideas. Attitudes expressed by mediums or even by the supposed deceased communicators that contradicted Spiritist doctrine were received with anxiety. Mediums tended to avoid doctrinal conflicts, attempting to maintain harmony and balance with both Spiritist and Christian values. This goal was not always accomplished owing to the difficulty of controlling the upswell from the unconscious that occurred during mediumistic trance. In such moments, while the medium's role was preserved, the spiritual harmony of the *Centro* was not. At those moments, the Spiritist center's concern with the maintenance of its Spiritist ideology overrode the development of the medium as an individual. Yet, even at moments when ideology was the main goal, the imposing of doctrine could lead to therapeutic and integrative outcomes. As Ciampa (1987) has argued, institutions must also undergo transformation and metamorphosis of their collective identity so as to adjust to the needs of their members and to the requirements of the social environment as a whole.

Maraldi also noted that the interviewees reported both some paranormal-like personal experiences that were better explained psychologically as the result of interpretation and cognitive process biases, group suggestion, or *criptomnésia*, and a mediumistic drawing that seemed to have resulted from a paranormal retrieval of information.[3]

From the perspective of mediumship as ideology, Maraldi had postulated that the medium's individual history seems to reproduce in several aspects the history of Spiritism in Brazil. For example, there is the fusion of a personal search with the still-unsolved collective search. Mediums' discourse—when it centered on debating scientific materialism and discussions about the supremacy of conventional Medicine and Catholic dogma, for instance—seem to reproduce (even indirectly) the early history of Brazilian Spiritism when suffering persecution by medical and religious institutions.[4] It seems that initial conflicts and continuing rivalries are alive in the mediums' imagination, keeping mediums on the defensive where their Spiritist beliefs are concerned. And this defensiveness, in fact, can be seen as an act of preservation of their own mediumistic identities.

Finally, Maraldi's study provides an important reflection based on I.N.'s life history which reveals a very significant development of psychologically adaptive identity that contradicts the idea that Spiritist beliefs and paranormal experiences indicate regression in the psychological sense as Ciampa and Habermas believe.

For Habermas (1990), paranormal beliefs are a regressive cultural form that represents a step backward in the process of identity development. Ciampa (1987) also believes that paranormal beliefs may be an obstacle to the achievement of Habermas's (1990) *post-conventional identity* as proposed, that is a relatively autonomous identity that moves beyond group and institution beliefs. Maraldi has noticed, however, that I.N. seems to be taking her first steps toward a *post-conventional* worldview, albeit without abandoning her paranormal beliefs. The formation of an adaptive identity has not cancelled her beliefs nor her commitment to her mediumship.

Developing a mediumistic worldview must be recognized as a valid way in which individuals may search for emotional and spiritual meaning in their lives, even if the outside observer doubts the reality of alleged Spiritist phenomena. A perspective that is both historical and psychological can help. Maraldi has proposed that the ideas presented by Foucault (1968) in *Mental Disease and Psychology* provide insight here. To Foucault, the regressive character of pathological behaviors in neurosis or in certain cultural phenomena—such as mystical and religious experiences—do not constitute an inherent expression of such phenomena, but rather reflect a culture that allocates to the past those elements that are offensive to the dominant worldview. The repression of such cultural phenomena would cause them to re-emerge as a kind of marginalized discourse. Linking the phenomena to pathology then can be seen as a cultural and historical construction, and not as an inherent property of the phenomena themselves. The pathologizing of mediumistic practices and Spiritist beliefs becomes, in a sense, more properly an effect rather than a cause.

So we would not agree that the majority of those who hold paranormal beliefs are members of socially marginalized groups such as women, the elderly, individuals of African descent, or homosexuals (Emmons & Sobal, 1981). Even if certain paranormal beliefs may be a psychological resource for dealing with the frustrations of social exclusion, such a generalization does not explain the enormous interest that members of socially privileged (dominant) groups have in the paranormal (Irwin, 2003). Perhaps such paranormal experiences as mediums claim and seek to develop can be seen as marginal not only because they are restricted to a specific segment of society or because they can serve as both a compensation and a justification for social and economic alienation in specific groups, but because they can be characterized as fundamentally resistant to the secularization of culture in its legitimate expression of a persistent spiritual aspect of the human condition.

Conclusion

In order to arrive at a more comprehensive understanding of mediumship, it is necessary to pay attention not only to its psycho–physiological aspects—as has been done in the past—but also to consider its psychosocial and cultural aspects. An exclusively psychopathological or intrapsychic interpretive axis has recently begun to transform into a psychosocial one, but it seems that the relationship of these perspectives to each other has not been successfully outlined as yet. It is not enough to point out both these perspectives as relevant. Rather it is necessary to determine *how* psychosocial aspects influence mediumistic experiences. This approach has inspired a number of recent studies (Machado, 2009, Maraldi, 2008, Negro, 1999, Stoll, 2004, Zangari, 2003), but more research is needed, especially that which can integrate the psychosocial with other aspects of mediumship.

We believe that the psychosocial perspective can promote a more effective scientific understanding of mediumship from that already provided by the theoretical heritage that reduces the social to the biological. It is necessary that different analytical perspectives—psychosocial and biological—be seen as complementary so that the complexity of mediums and mediumship may be better understood.

Notes

[1] Allan Kardec was the name adopted by the French mathematician and educator Hyppolite Leon Denizard Rivail (1804–1869), who was the codifier of Spiritism. In the second part of 19th century, he was invited to analyze and organize several reports received by mediums working in different Spiritist groups—especially those linked to the historical tradition of belief in survival and to the principle that the deceased can interact in our world (Kardec, 1861/2001). In Brazil, Kardec's Spiritist perspective—Kardecism—is the most influential of the Spiritist traditions and has more followers than other similar religions.

[2] Other societies or institutes with similar objectives were later founded, such as the *American Society for Psychical Research* (1885) in the USA and the *Institut Métapsychique International* (1919) in Paris, for example.

[3] This is discussed in more detail in Maraldi's (2008) exploratory study. However, it was not possible to get a broad comprehension of the nature of the "secondary personalities" manifested by the interviewed mediums.

[4] Nowadays people in Brazil are free to adhere to any religion and to express their faith. The country is said to be the biggest Spiritist country in the world. There are Kardecist Spiritist centers as well as other mediumistic religious temples such as Umbanda and Candomblé all over the country. At the same time, Brazil is one of the biggest Catholic countries in the world. In fact, it is common to find Brazilians who are adept at more than one religion, especially Catholicism and Kardecist Spiritism (Machado, 2009).

Acknowledgements

Thank you to CNPq, Conselho Nacional de Desenvolvimento Científico e Tecnológico (Brazil), for the grant received by Everton de Oliveira Maraldi for his Master's Degree research, to Carlos S. Alvarado for his help and suggestions, and especially to Nancy L. Zingrone for editing this paper in English.

References

Almeida, A. M. (2004). *Fenomenologia das experiências mediúnicas: Perfil e psicopatologia de médiuns espíritas*. [Ph.D. Dissertation]. Faculdade de Medicina. São Paulo: Universidade de São Paulo.
Almeida, A. M., Almeida, A. A. S., & Lotufo Neto, F. (2005). History of "Spiritist madness" in Brazil. *History of Psychiatry, 16*(1), 5–25.
Almeida, A. M., Lotufo Neto, F. (2004). A mediunidade vista por alguns pioneiros da área mental. *Revista de Psiquiatria Clínica, 31*(3), 132–141.
Almeida, A. M., Lotufo Neto, F., & Greyson, B. (2007). Dissociative and psychotic experiences in Brazilian Spiritist mediums. *Psychotherapy and Psychosomatics, 76,* 57–58.
Alvarado, C. S. (2002). Dissociation in Britain during the late nineteenth century: The Society for Psychical Research, 1882–1900. *Journal of Trauma and Dissociation, 3*(2), 9–33.
Alvarado, C. S. (2005). Historical notes on the role of mediumship in Spiritualism, Psychical Research and Psychology. *The Parapsychology Foundation Conference: The study of Mediumship: Interdisciplinary Perspectives*. Charlottesville: Virginia. (Abstract available at http://www.pflyccum.org/264.html Accessed 14 October 2009)
Alvarado, C. S., Machado, F. R., Zingrone, N., & Zangari, W. (2007). Perspectivas históricas da influência da mediunidade na construção de idéias psicológicas e psiquiátricas. *Revista de Psiquiatria Clínica, 34*(1), 42–53.
Augras, M. (1983). *O Duplo e a Metamorfose*. Petrópolis: Vozes.
Bastide, R. (1989). *As Religiões Africanas no Brasil: Contribuição a uma Sociologia das Interpenetrações de civilizações*. São Paulo: Pioneira. (Original work published in 1960)
Braude, S. E. (1988). Mediumship and multiple personality. *The Journal of the American Society for Psychical Research, 55*(813), 177–195.
Ciampa, A. C. (1987). *A estória do Severino e a história da Severina: Um ensaio de Psicologia Social*. São Paulo: Brasiliense.
Ciampa, A. C. (1994). Identidade. In: Lane, S. T. M., & Codo, W. (Eds.), *Psicologia social: O homem em movimento*. São Paulo: Brasiliense.
Costa, Jurandir Freire (1976). *História da Psiquiatria no Brasil*. Rio de Janeiro: Documentário Cruz Monclova Lidio.
Ellenberger, H. F. (1976). *El Descobrimento del Inconsciente*. Madrid: Gredos.
Emmons, C. F, & Sobal, J. (1981). Paranormal beliefs: Testing the marginality hypothesis. *Sociological Focus, 14,* 49–56.
Flournoy, T. (2007). *Spiritism and Psychology*. New York: Cosimo Classics. (Original work published in 1911)
Foucault, M. (1968). *Doença mental e Psicologia* (L. R. Shalders, trans.). Rio de Janeiro: Tempo Brasileiro (Biblioteca Tempo Universitário, Volume 2).
Giumbeli, E. (1997). Heresia, doença, crime ou religião: O Espiritismo no discurso de médicos e cientistas sociais. *Revista de Antropologia, 40*(2), 31–82.
Giumbeli, E. (2003). O "Baixo Espiritismo" e a História dos Cultos Mediúnicos. *Horizontes Antropológicos, 9*(19), 247–281.
Grosso, M. (1997). Inspiration, Mediumship, Surrealism: The Concept of Creative Dissociation. In: Krippner, S., & Powers, S. M. (Eds.), *Broken Images, Broken Selves: Dissociative Narratives in Clinical Practice*. Washington, D. C.: Brunner/Mazel.
Habermas, J. (1990). *Para a Reconstrução do Materialismo Histórico* (C. N. Coutinho, trans.). São Paulo: Brasiliense.
Herskovits, M. J. (1967). *Les bases de l'anthropologie culturelle*. Paris: Payot.
Hess, D. (1991). *Spiritists and Scientists. Ideology, Spiritism, and Brazilian Culture*. Pennsylvania: The Pennsylvania State University Press.
Hughes, D. J. (1992). Differences between trance channeling and multiple personality disorder on structured interview. *The Journal of Transpersonal Psychology, 2,* 181–192.
Irwin, H. J. (2003). *An Introduction to Parapsychology, Fourth Edition*. Jefferson: McFarland.
Jung, C. G. (1920/2004). *O Eu e o Inconsciente* (Dora F. da Silva, trans.). Petrópolis: Vozes, 2004.

(*Obras Completas de Carl Gustav Jung*, volume 7)
Kardec, A. (2001). *O Livro dos Médiuns: Guia dos Médiuns e dos Doutrinadores* (J. Herculano Pires, trans.). São Paulo: Lake. (Original work published in 1861)
Le Maléfan, P. (1999). *Folie et Spiritisme: Histoire du Discourse Psychopathologique sur la Pratique du Spiritisme, ses Abords et ses Avatars (1850–1950)*. Paris: L'Hartmattan.
Machado, F. R. (1996). *A Causa dos Espíritos: Estudo sobre a Utilização da Parapsicologia para a Defesa da Fé Católica e Espírita no Brasil*. [Master's thesis]. Ciências da Religião. Pontifícia Universidade Católica de São Paulo.
Machado, F. R. (2005). Parapsicologia no Brasil: Entre a Cruz e a Mesa Branca. *Boletim Virtual de Pesquisa Psi, 2*. http://www.pesquisapsi.com.br
Machado, F. R. (2009). *Experiências Anômalas na Vida Cotidiana: Experiências Extra-Sensório-Motoras e sua Associação com Crenças, Atitudes e Bem-Estar Subjetivo*. [Tese (Doutorado)]. [Doctoral thesis]. Instituto de Psicologia: Universidade de São Paulo.
Maraldi, E. O. (2008). *Um Estudo Exploratório Sobre os Usos e Sentidos das Crenças e Experiências Paranormais na Construção da Identidade de Médiuns Espíritas*. [Monograph]. São Paulo: Curso de Psicologia, Universidade Guarulhos.
Myers, F. W. H. (2001). *Human Personality and Its Survival of Bodily Death*. Charlottesville, VA: Hampton Roads Publishing Company Inc. (Original work published in 1903)
Negro, P. J. (1999). *A Natureza da Dissociação: Um Estudo sobre Experiências Dissociativas Associadas a Práticas Religiosas*. Tese (Doutorado) [Doctoral thesis]. São Paulo: Faculdade de Medicina, Universidade São Paulo.
Nina Rodrigues, R. (1900). *L'Animisme Fétichiste des Nègres de Bahia*. Salvador: Reis.
Northcote, J. (2007). *The Paranormal and the Politics of Truth: A Sociological Account*. Exeter, UK: Imprint-Academic.
Oliveira, A. X. de (1931). *Espiritismo e Loucura*. Rio de Janeiro: A. Coelho Branco.
Owen, A. (1990). *The Darkened Room: Women, Power and Spiritualism in Late Victorian England*. Philadelphia: University of Pennsylvania Press.
Querino, M. (1955). *A Raça Africana*. Salvador: Livraria Progresso.
Reinsel, R. (2003). Dissociation and Mental Health in Mediums and Sensitives: A Pilot Survey. *Proceedings of Presented Papers* (pp. 200–221). (The Parapsychological Association 46th Annual Convention; Vancouver BC; August 2–4, 2003)
Ribeiro, R. (1978). *Cultos Afro–Brasileiros do Recife*. Recife: I.J.N.P.S.
Ribeiro, L., & Campos, M. (1931). *O Espiritismo no Brasil. Contribuição ao Seu Estudo Clínico e Médico–Legal*. São Paulo: Editora Nacional.
Richeport, M. M. (1992). The interface between multiple personality, spirit mediumship and hypnosis. *American Journal of Clinical Hypnosis, 34*(3), 168–177.
Rogo, D. S. (1986). *Life after Death: The Case for Survival of Bodily Death*. London: Guild.
Ronan, C. A. (2001). *História Ilustrada da Ciência da Universidade de Cambridge: Da Renascença à Revolução Científica* Volume 3 (J. E. Fortes, trans.). Rio de Janeiro: Jorge Zahar.
Shamdasani, S. (1994). Encountering Hélène: Théodore Flournoy and the Genesis of Subliminal Psychology. In: Flournoy, T. (1900), *From India to the Planet Mars: A Case of Multiple Personality with Imaginary Languages,* Princeton, NJ: Princeton University Press, pp. xi–xii.
Stoll, S. J. (2004). Narrativas Biográficas: A Construção da Identidade Espírita no Brasil e sua Fragmentação. *Estudos Avançados, 18*(52), 181–199.
Verger, P. (2002). *Saída de IAO—Cinco Ensaios sobre a Religião dos Orixás*. São Paulo: Axis Mundi.
Zangari, W. (2003). *Incorporando Papéis: Uma Leitura Psicossocial do Fenômeno da Mediunidade de Incorporação em Médiuns de Umbanda*. Tese (Doutorado) [Doctoral thesis]. São Paulo: Instituto de Psicologia, Universidade de São Paulo.
Zingrone, N. L. (1994). Images of Woman as Medium: Power, Pathology and Passivity in the Writings of Frederic Marvin and Cesare Lombroso. In: Coly, L., White, R. A. (Eds.), *Women and Parapsychology: Proceedings of an International Conference* (pp. 90–123). New York: Parapsychology Foundation,

ESSAY

Investigating Mental Mediums:
Research Suggestions from the Historical Literature

CARLOS S. ALVARADO

Atlantic University, 215 67th Street, Virginia Beach, Virginia 23451, USA
carlos.alvarado@atlanticuniv.edu

Abstract—Mental mediumship is a complex process involving a variety of factors in need of further study before we can increase our understanding of the phenomenon. The purpose of this paper is to offer ideas and topics for further research—mainly from the psychological perspective and with emphasis on the old psychical research literature. The topics discussed are mediumistic trance (e.g., function, stages, and depth, mediumistic mentation (e.g., imagery, symbols), the dramatic capabilities of the subconscious mind, the relationship between mediumship and psychopathology, the variety of experiences reported by mediums outside their performances (e.g., dissociative and ESP experiences), and the changing aspects of mediumship over time. It is argued that in-depth single case studies of specific mediums and interdisciplinary studies will greatly help us to understand mediumship more fully.

Keywords: mediumship—mediumistic mentation—personation—iatrogenesis
—trance—dissociation

Mental mediums are individuals who claim to convey messages from discarnate spirits in such varied ways as impressions, visions, and automatic writing. Historically, this phenomenon has been important for its influence on psychological concepts such as the subconscious mind and dissociation (Alvarado, Machado, Zangari, & Zingrone, 2007), ideas of pathology (Le Maléfan, 1999), and for presenting phenomena that have provided both research topics and ideas for fields such as spiritualism (Tromp, 2006), psychical research (Inglis, 1984), and anthropology (Seligman, 2005). Modern studies of the phenomena have included approaches and questions framed from such varied fields as anthropology, history, parapsychology, psychology, and psychiatry, to name a few. Ideally, to understand mediumship we need a comprehensive approach that considers multiple variables and the contribution of different disciplines. While practical considerations limit such research in a single project, perhaps we might follow more general approaches within specific disciplines. For example, Frederic W. H. Myers (1903) not only focused on veridical mediumistic communications in his analysis of mediumship, but he also took a much wider scope considering non-veridical manifestations as well as non-mediumistic motor automatisms and secondary personalities.

The purpose of this paper is to suggest some topics for further research with mental mediums, taking ideas mainly from the old psychical research literature, and, to some extent, from the spiritualist and psychological literatures. Because of the type of literature consulted, the examples cited are mainly on trance mediums. But the systematic study of mediums can and should also include mediums who do not fall into trance, a type that seems to be more frequent today. Although the topic will not be completely neglected, my emphasis will not be on studies of veridical mediumship (e.g., Beischel & Schwartz, 2007, Hodgson, 1892). Instead, I will focus on a variety of questions related to trances, mediumistic mentation, the dramatic capabilities of the subconscious, psychopathology, the variety of cognitive and psychic experiences reported by mediums in their lives, and changes in mediumship over time. Rather than present detailed suggested research designs or reviews of past studies, my intent is to raise questions that may inform current hypothesis testing.

Trance

The term *trance*[1] is a problematic one. Not only has it been used to refer to a variety of apparent states of consciousness, but we need to be aware that it may manifest in degrees, a topic that has been discussed in the psychical research (Sidgwick, 1915) and spirit possession (Frigerio, 1989) literatures. In his influential study of music and trance, Rouget (1980/1985:3) pointed out that some individuals who have referred to trance have not used the same terms to designate the manifestations and that authors have used different terms to refer to the same phenomena. Pekala and Kumar (2000) have argued in relation to hypnosis that the concept of trance is ill-defined and has not been properly operationalized. The situation is similar in the mediumship literature. Nonetheless, while we need to keep this problem in mind, I will use the term *trance* here because the word is used to describe apparent changes in states of consciousness in the literature I am reviewing.

Trance has long been considered important for the manifestation of phenomena. According to James H. Hyslop (1918): "The trance of the living medium more or less excludes her own mind or thoughts from intermingling with or dominating the messages . . . ." (p. 218). Unfortunately, there has been no systematic research on the subject to test such an idea. Perhaps a contrast of the medium's experiences in trance vs. no trance, or between different stages of trance, could produce information relevant to the subject.

Medium Leonora E. Piper illustrates interesting research possibilities. As described by Charles Richet, Piper started in silence and in semi-darkness and, after an interval of 5 to 15 minutes, exhibited "small spasmodic convulsions" that ended in a small epileptic episode (in W. Leaf, 1890:619). [This and other translations in this paper are mine.] Following this, the medium showed stupor and heavy breathing, and then started talking. Others observed similar convulsive behaviors with Piper, generally at the beginning of the trance: "She twitched convulsively, ejaculated 'don't,' and went into apparent epilepsy" (Lodge, 1890:444); "She continually groaned as if in suffering. After long waiting . . . she went through a kind of struggle or crisis, confined to the upper part of the body . . . ." (W. Leaf, 1890:606); "Convulsions strong; continue ten minutes" (Hodgson, 1892:537). There were similar observations at the end of the trance: "Mrs. Piper had begun to come out of trance, but was strongly convulsed again . . . ." (Hodgson, 1892:483).[2]

While not every medium shows these manifestations, it would be worth determining how prevalent they are. Psychophysiological recordings at the time the tremors take place may be compared with periods without the tremors. It seems these phenomena were more frequent during the early years of Piper's trances. Are such manifestations developmental? That is, did they appear during the early years of the medium's practice as her nervous system developed or was trained (so to speak) to manifest the phenomena, disappearing as the processes involved became second nature?

Eleanor Sidgwick (1915) documented the existence of stages of trance and the variety of phenomena accompanying Mrs. Piper's mentation. Particularly interesting were Hodgson's (1898) observations of an initial stage in which he believed the subliminal mind of the medium took over at the beginning of the trance in which the medium was "dreamily conscious of the sitter, and dreamily conscious of 'spirits'" (Hodgson, 1898:397). This led to a "fuller and clearer consciousness—we may call it her subliminal consciousness—which is in direct relation . . . not so much with our ordinary physical world as with 'another world'" (p. 397), and later into a state in which he thought the "subliminal consciousness withdraws completely from the control of her body and takes her supraliminal consciousness with it" (p. 398). Hodgson wrote further that at the end there were indications of a return to Mrs. Piper's consciousness in reverse of the order it had disappeared. He wrote:

> But in passing out of trance, the stages are usually of longer duration than when she enters it. She frequently repeats statements apparently made to her by the "communicators" while she is in the purely "subliminal" stage, as though she was a "spirit" controlling her body but not in full possession of it, and, after her supraliminal consciousness has begun to surge up into view, she frequently has visions apparently of the distant or departing "communicators". (Hodgson, 1898:400–401)

Unfortunately, very little work has been conducted since the old days to assess the existence and characteristics of stages of mediumistic trance. Many basic questions need to be explored empirically. In addition to the abovementioned—the actual function of trance, its psychophysiology, features, and stages—there is the issue of veridical mediumistic mentation. Just as modern parapsychologists have studied ESP in relation to aspects of altered states of consciousness (e.g., depth, changes in sense of time and body image (Alvarado, 1998)), we may study veridical mentation in mediums in relation to the presence or absence of trance and its depth.

While trance mediums do not seem to be as common today as they used to be, observations of such states suggest that psychophysiological recording techniques could be profitably used in new studies, as has been done sporadically in the past (Evans & Osborn, 1952, Solfvin, Roll & Kelly, 1977). On a more basic level, we may better document today the variety of states of consciousness shown by some mediums, such as was the case with Mrs. Willett (pseudonym of Winifred Coombe-Tenannt (Balfour, 1935)).

Mediumistic Mentation

There are discussions in the literature about the features and difficulties of both mediumistic mentation and the apparent process of communication assuming the action of spirits (e.g., Hodgson, 1898, Hyslop, 1919). For example, Hyslop (1919) referred to the "pictographic process," in which "the communicator manages to elicit in the living subject a sensory phantasm of his thoughts, representing, but not necessarily directly corresponding to, the reality" (Hyslop, 1919:111). Some messages were expressed in visual, or in other, modalities, according to the sensitivity of the medium. Motor expressions of mediumship, such as automatic writing, do "not represent anything pictographic, though pictographic processes may precede it" (Hyslop, 1919:111). Regardless of Hyslop's emphasis on discarnate agency, in his view there could be confusions and distortion in the mentation due to the interpretation of images involving the subconscious mind of the medium. Such an idea opens possibilities for research to explore intrapsychic factors involved in mediumistic mentation. Among other aspects, such studies could use the various ways developed to study imagery, nonverbal abilities, and preferred modes of mentation (Mammarella, Pazzaglia & Cornoldi, 2006, Richardson, 2006, Riding, 2006).

Similarly, new studies could consider difficulties in expressing messages that may be caused by memory retrieval problems and other factors. One line of research that might be relevant is that concerning tip-of-the tongue states in which something that a person knows cannot be recalled (Schwartz, 2002).

Another topic deserving study, and one discussed from the early days of spiritualism, is the influence of the medium on the communications. An early American researcher and medium, John W. Edmonds, wrote: "I know of no mode of spiritual intercourse that is exempt from a mortal taint—no kind of mediumship where the communication may not be affected by the mind of the instrument" (Edmonds & Dexter, 1855(2):39). Similarly, another author stated: "All communications partake more or less of the character of the media [mediums] through whom they pass" (Barkas, 1862:102).

Years later, Myers (1902) wrote about the mediumship of Rosina Thompson and suggested that confusions and mistakes may come "mainly from Mrs. Thompson's own subliminal self" (Myers, 1902:72). Myers (1903), who discussed both the influences of discarnate influence and the influence of the medium, also referred to the mix of both sources. He also commented on "the influence of the sensitive's supraliminal self . . . whose habits of thought and turns of speech must needs appear whenever use is made of the brain-centres which that supraliminal self habitually controls" (Myers, 1903(2):249).[3] Content analyses of mediumistic mentation may be helpful in identifying the medium's waking memories, turns of phrase, and other idiosyncrasies in their mentation.

Similarly, content analysis may help us to identify and classify symbols in mediumistic productions. Several writers have presented examples of symbols in mediumistic communications (Bozzano, 1907:253–254, Emmons & Emmons, 2003, Chapter 69). According to Hereward Carrington (1920): "It is in the interpretation of these symbols that much of the true art of mediumship and psychic development will be found to lie . . ." (p. 109), but he added that each medium "must learn . . . , by repeated experience, what the various symbols mean . . . , and thus form a 'code' or method of interpretation . . . . " (p. 109). Consistent with this, and based on their analyses of the experiences of many mediums, Emmons and Emmons (2003) stated that "to a great extent mediums have separate 'psychic dictionaries'" (p. 258).

Saltmarsh (1929) explored symbols in his study of Mrs. Warren Elliott. The medium received messages from a control called Topsy such as the following: "'You nearly married and then not married. Shows Topsy like wedding dress and then sort of drops it.' (Symbol is obvious here.)" (p. 121). Saltmarsh wrote:

> It will be observed that the symbols are all of a certain type. They are what might be called natural symbols, and are based on habitual analogies, either verbal, as for example when the hallucinatory figure coming near to the sitter is taken to mean nearness of relationship, or common forms of speech, as when all black is used as a symbol for worry or sorrow; or else they may be natural pantomime, as when the gesture of waving away is interpreted as meaning that the ostensible communicator was not connected with the relic. (p. 123)

A study could be designed in which symbols appearing in the mentation of particular mediums are classified by type and then compared both between and within mediums.[4] While we may find some similarities or consistency, there is likely to be much symbolism that is particular to specific mediums. The issue, of course, is an empirical one and should be studied taking into consideration variables that may affect the symbols. This includes the general spiritualist beliefs and interests specific to circles in which the medium was trained or worked during his or her life. There is also much to investigate regarding the factors that may affect both the formation and the manifestation of symbols.

Regardless of common patterns, there are probably many individual differences in the content of mediumistic communications. But even if this is not the case, we could learn much about aspects of the mentation, such as the flow of imagery, and salient features such as the repeated use of specific images in relation to topics and communicators, as well as veridical messages.

The latter was documented by Charles Drayton Thomas (1928, 1939) in his study of Gladys Osborne Leonard. Communications from the medium's control Feda frequently alluded to seeing letters instead of perceiving names or places. The following examples of Feda's statements, published by Thomas (1939), are followed by explanations of the possible meaning of the letters in parentheses. "A place beginning with the letter 'S,' very much connected with him and his work. A place with a good many letters in its name. (When he came home from the West Indies he and his family lived in Southampton for many years)" (Thomas, 1939:261). Another example read as follows: "Was this Mr Macaulay a researcher of some kind? . . . He always wanted fuller information. He read much about it and went somewhere where he could study this "M" subject. ("M" is possibly a reference to Meteorology)" (Thomas, 1939:273).

Obviously not all mediums have this type of mentation with letters. A comparison of mediums with this imagery style with those who have other styles may help us to assess, for example, if such manifestations are related to cognitive differences, or to differences in training.

There are other examples with Mrs. Leonard of the features of reception of messages, and difficulties in expressing them. Troubridge (1922) referred to her observation that information sometimes came from the medium "bit by bit, conveying to the recipient or listener the impression of someone finding isolated pieces all connected with one jigsaw puzzle and dealing them out one by one with a view to their being fitted together by the recipient into an intelligible whole" (Troubridge, 1922:371).[5] This is evident in an incident reported by Radclyffe-Hall and Troubridge (1919). One of the communicators, when alive, used the word "spork" and "sporkish" to refer to unpleasant people. In a seance held on January 17, 1917, Feda, Leonard's spirit control, was having trouble getting a word. She said: "Feda can't get it. But it is only a short word" (Radclyffe-Hall

& Troubridge, 1919:445). She traced the letter "S" on the palm of a sitter. At a sitting on May 2, the word still could not be obtained "beyond making the opening sibilant consonant" (Radclyffe-Hall & Troubridge, 1919:446). Feda said in a low voice: "It's *what*, Ladye? What are you trying to say S-ss-Sss-S-ss . . . . What *is* the word, Ladye? It's Spor-Spor-Spor!" (Radclyffe-Hall & Troubridge, 1919:446). More faltering attempts followed, and she said "Sporti" and "Sporbi." The account continued as follows:

> After this small letter there comes a curved letter, and then it seems to Feda there's another long letter . . . It's S . . . P . . . O . . . then a little letter and then a letter like this; (she draws a "k" . . .). It's a down stroke like this, with a little bit like this sticking on to it; Sporki . . . Sporkif? . . . . That letter goes like this . . . (here Feda draws an "S" . . .). And then there's another letter like this (here Feda draws an "H"). . . . Well, Feda can't see any more, (suddenly and very loud) SPORKISH! SPORKISH! But that isn't a word at all! . . . "Sporkish," she says it in such a funny way . . . she says that you and she used to call people that sometimes. . . . (Radclyffe-Hall & Troubridge, 1919:447)

In addition to the imagery and fragmentary nature of many communications, we could focus new studies on the variety of ways mediums manifest their communications. For example, Mrs. Willett's mediumship consisted of impressions of presences, mental images, feelings and emotions, impulses and inhibitions, and verbal images (Balfour, 1935). Modern research could try to assess how many mediums combine these and other forms of expression, as well as how they may specialize almost completely in particular forms of expression.

Some modern researchers have analyzed aspects of the content of mediumistic mentation (K. W. Barrett, 1996, Emmons & Emmons, 2003, Rock, Beischel, & Schwartz, 2008). For example, Rock et al. (2008) studied a small number of mediums in terms of aspects such as the sensory modalities they experienced, feelings about ailments or cause of death, and changes of affect. I hope work along this line continues following up on the above-mentioned observations so that we may be able to replicate and extend previous findings. Perhaps some of the scales used by Pekala (1991)—including questions about such varied aspects as changes in perception and the sense of time, as well as imagery vividness and positive and negative affect—may be used with mediums, or adapted for such work even considering the problems of using such instruments to quantify experiences (Stevens, 2000).

The Dramatic Capabilities of the Subconscious

From the early days of spiritualism, some authors have mentioned that mediums may have the potential of dramatizing changes of personality. One writer wrote that in some circumstances the medium's automatic brain functions could "assume any personality, from that of a divinity to that of a toad . . . " (Rogers, 1853:171). Another suggested that some mediumistic communications about the nature of the other world could "proceed from the poetic brains of the writers, and are not the product of any disembodied spiritual intelligences" (Barkas, 1862:134).

Myers (1884, 1885) argued early on for the interpretation of some mediumship on the basis of subconscious creations by the living mind of the medium, and sometimes transcending intrapsychic processes through recourse to telepathy. Eduard von Hartmann (1885) wrote about a somnambulistic consciousness in mediums that "inclines to symbolising and personifications," showing a dramatic "metamorphosising talent" (p. 453, both quotes) to produce fictitious communications. He postulated that such consciousness could obtain information from the waking consciousness and memories of the medium, as well as through telepathy and clairvoyance.

Many referred to the potential of the subconscious to dramatize spirit communications. Myers (1903(2):130) wrote about a "strange manufacture of inward romances." Théodore Flournoy (1900:425) studied what he described as a "tendency of the subliminal imagination to reconstruct the deceased and to feign their presence" in some mediumistic communications, as well as to a "spirit-imitating" process existing within us (Flournoy, 1911:202). In his *Traité de Métapsychique*, Charles Richet (1922) discussed what he called the "talents of the unconscious" (p. 50), or capabilities such as having a more detailed memory than the conscious mind. This and other abilities, he stated, allowed for the possibility of producing all kinds of phenomena, including those presented by mediums pretending to be the results of spirit influence.

An important concept to understand such tendencies is the view that the production of fictitious communicators and stories in mediumistic mentation may be a function of unintended suggestion, expectations, and beliefs in the social and psychological surroundings of the medium. The idea was applied during the nineteenth century and later to hypnotic manifestations and to such dissociative phenomena as secondary personalities (Alvarado, 1991). Some hypnotic phenomena, Delboeuf (1886) suggested, were due to the influence of imitation and education. Referring to suggestion, Pierre Janet in *L'Automatisme Psychologique* (1889) commented about secondary personalities in hypnotic subjects: Once named, "the unconscious personage is more determined and more distinct, it shows better its psychological characteristics" (p. 318).

Spiritualists also referred to similar ideas when discussing mediumistic communications. Some of these individuals were skeptical about the validity of communications about reincarnation based on the teachings of the "spirits" presented by Allan Kardec (e.g., 1867, Part 2, Chapters 4–5). One such critic was Alexandre Aksakof (1875). In his critique of Kardec, he referred to writing mediums who "pass so easily under the psychological influence of preconceived ideas. . . ." (Aksakof, 1875:75). An anonymous author emphasized that the few English mediumistic discussions of reincarnation had been "strongly coloured by the opinions of the medium, or those of the sitter. . . ." ("Allan Kardec's 'Spirits' Book, 1875:170). In his view, if reincarnation received more attention in England, "plenty of spirits will begin to teach it, the reason being that the minds of the various mediums will be set buzzing by the arguments on the subject mooted by persons around them. . . ." ("Allan Kardec's 'Spirits' Book", 175:170).[6]

In *The Principles of Psychology*, William James (1890) emphasized that the Zeitgeist influenced mediumistic productions. He wrote:

> One curious thing about trance-utterances is their generic similarity in different individuals. The "control" here in America is either a grotesque, slangy, and flippant personage ("Indian" controls, calling the ladies "squaws", the men "braves", the house a "wigwam", etc., are excessively common); or, if he ventures on higher intellectual flights, he abounds in a curiously vague optimistic philosophy-and-water, in which phrases about spirit, harmony, beauty, law, progression, development, etc., keep recurring. It seems exactly as if one author composed more than half of the trance-messages, no matter by whom they are uttered. Whether all sub-conscious selves are peculiarly susceptible to a certain stratum of the *Zeitgeist*, and get their inspiration from it, I know not; but this is obviously the case with the secondary selves which become "developed" in spiritualist circles. There the beginnings of the medium trance are indistinguishable from effects of hypnotic suggestion. The subject assumes the rôle of a medium simply because opinion expects it of him under the conditions which are present; and carries it out with a feebleness or a vivacity proportionate to his histrionic gifts. (James, 1890(1):394)[7]

The writings and studies of Théodore Flournoy (1900, 1901, 1911) are of key importance for the subject. He studied communications in which the medium Hélène Smith described previous existences in India, France, and life on planet Mars, including the creation of a Martian language. As he wrote:

> We must . . . take into consideration the enormous suggestibility and auto-suggestibility of mediums, which render them so sensitive to all the influences of spiritistic reunions, and are so favorable to the play of those brilliant subliminal creations in which, occasionally, the doctrinal ideas of the surrounding environment are reflected together with the latent emotional tendencies of the medium herself. (Flournoy, 1900:443)

In addition to suggestibility, we could consider the concept of emotional contagion that has a long tradition in psychology (Levy & Nail, 1993). The concept has been related to hypnotic experience (Cardeña, Terhume, Lööf, & Buratti, 2009).

Maxwell (1903/1905) used the term personification to refer to the intelligence behind mediumistic phenomena, an intelligence he believed came from the medium. The identity of the personification changed according to the individuals in the mediumistic circle:

> I have noticed that the role played by the personification varies with the composition of the circle. It will always be the spirit of a dead or living person with spiritists. But the roles are more varied if the circle is composed of people who are not spiritists; it then sometimes happens that the communications claim to emanate from the sitters themselves. (Maxwell, 1903/1905:65)

The personification, Maxwell believed, was very suggestive.

Many authors discussed the importance of suggestions and beliefs on the development of trance personalities and the stories of the communicators, among them Pierre Lebiedzinski (1924), Amy Tanner (1910), and Réne Sudre (1926). The latter discussed what he termed "prosopopesis," or "brusque, spontaneous or provoked changes of psychological personality" (Sudre, 1926:85) developed by the subconscious mind in mediumship, as well as in possession, multiple personality, and under the influence of hypnosis.[8] Eleanor Sidgwick (1915) commented about this topic:

> That the sitters must influence the trance communications to some extent is . . . obvious. For one thing, they are themselves personages in the drama, and the part they play in it and the way they play it must affect the way the trance personalities play theirs. . . . And in the trance drama the sitters not only largely determine the subjects of conversation, but the personages who shall take part in it. They explicitly or tacitly demand that their own friends shall manifest themselves and produce evidence of identity, or give information on particular points. (Sidgwick, 1915:294)

Sidgwick's point is well-taken. Such influence may be seen in analyses of the verbal interactions between mediums and their sitters (Wooffitt, 2006).

Communications about life on other planets such as those discussed by Flournoy (1900, 1901) are good examples of possible iatrogenic creations. But there are several other examples in the literature about interplanetary communications (e.g., Hyslop, 1906, Jung, 1902/1983:34–35, Sardou, 1858, Weiss, 1905). Astronomer Camille Flammarion (1907) explained a particular case in which there were mediumistic drawings of houses on Jupiter (Sardou, 1858) as "the reflex of the general ideas in the air" (p. 26). New cases of this sort could be studied to attempt to determine the specific influences leading to such productions.

Interestingly, some mediums known for producing veridical mentation have also presented non-veridical material that may be the result of the demands and suggestions provided by researchers and sitters. A case in point was Mrs. Leonard. In Thomas' book *Life Beyond Death with Evidence* (1928), there are chapters about evidential and non-evidential communications. The latter included descriptions of life in the spirit world, and ideas about the process of mediumistic communication.[9] The impression I get going over the book is that Mrs. Leonard had the ability to produce material about almost every topic.[10] While this may be consistent with the idea that mediumistic manifestations may be shaped by the interests of individuals around mediums, it is not proof that the medium was suggestible. In any case, we need to be more active in exploring the plasticity of the medium's psychological resources, and the limits of such processes.

One way to test for the influence of ideas on mediumistic mentation is to induce particular ideas through direct suggestions, as Richet (1883) and Harriman (1942) did in the creation of secondary personalities with research participants who were not mediums. This brings to mind the famous case in which G. Stanley Hall created a fictitious "Bessie Beals" personality with Mrs. Piper. The personality was presented to Mrs. Piper as his niece and the suggestion influenced the medium (Tanner, 1910:176, 181, 195–196, 254).[11]

However, indirect methods such as creating a context with conversations and information about what is desired may also be successful. Mediumship is probably affected to some extent by the intellectual and psychological environment around the medium, not to mention by the demands of the sitters (e.g., Lebiedzinski, 1924). The desire of members in mediumistic circles, or of individuals consulting the medium, to obtain philosophical communications about life or about the afterlife may not only shape but also train the medium to produce such manifestations. Similarly, interest in such aspects as evidential communications may lead the phenomena in that direction. Although, in theory, some mediums may be "fixed" into phenomena or topics in communications, they may also change according to the circle. An interesting study would be one in which mediums known for their production of non-evidential phenomena are influenced both directly and indirectly to produce veridical communications.

But regardless of what some have suggested (e.g., Tanner, 1910), the situation is not that simple. A medium is not a parrot at our command who can be shaped without limits. First of all, as in hypnotic suggestion, there are bound to be individual differences in the suggestibility of mediums. Second, we may find that, regardless of past discussions (e.g., Flournoy, 1900, Tanner, 1910), such effects are more difficult to produce than previously assumed. Nonetheless, research on the topic may allow us to empirically assess ideas about the plasticity of the medium's mind.

While in theory it makes sense to study this topic through some sort of manipulation, in practice there are ethical problems. Such an induction of phenomena would involve deception and may go against the medium's beliefs. Considering such objections, perhaps we would do well to see if studies similar those of Flournoy (1900), in which no manipulations were performed, could be conducted with modern mediums.

Psychopathology

Another research topic is the often-discussed issue of a relationship between mediumship and psychopathology, a topic discussed frequently during the nineteenth century. For example, in 1860 an anonymous writer in the medical journal *Lancet* argued that the "counterpart of the wretched medium we find in the half-deluded and half-designing hysterical patient" (*The delusion of spiritualism,* 1860:466). Similar thoughts were presented by many physicians who believed mediumship was pathological (for overviews, see Alvarado et al., 2007:48–50, Le Maléfan, 1999, Moreira-Almeida, Almeida, & Lotufo Neto, 2005).

The French were particularly interested in these issues. Pierre Janet (1889) said that mediums had frequent nervous "accidents" that included convulsions, choreic movements, and nervous crises. In his view, mediumship "depends on a particular morbid state" (Janet, 1889:406), similar to what later may develop as hysteria or insanity. But he argued that "mediumship is a symptom and not a cause" (Janet, 1889:406). Years later Janet (1909) presented a case of a 37 year-old woman who produced automatic writing messages. Most of the writings were said to come from her deceased father and were about the woman's clothing and hygiene. Janet believed that the woman's constant communications represented a case of "systematic" delusion.

"From time to time," wrote Alfred Binet in his book *Les Altérations de la Personnalité* (1892), "the most discreet authors cannot avoid saying that such an excellent medium has had a nervous crisis or gets fatigued quickly as the result of a too delicate health. . . . " (p. 299). J. Lévy-Valensi (1910) argued that the mental "disaggregation" (dissociation) of mediums became habitual, resulting in a delusion. Such condition could include hallucinations, erotic sensations, and problems with genital functions, and defensive reactions such as rituals and practices. He mentioned 17 cases of mediums of which six were said to have become insane due to their practices. Another French physician, Gilbert Ballet (1913), created a mediumistic diagnosis called Chronic Hallucinatory Psychosis that had the "disaggregation" of personality as the main symptom, as well as delusions of persecution and ambition, and hallucinations.

A problem with many of these observations is that they seem to have been done with clinical patients instead of with practicing mediums. Anyone who knows the careers of mediums such as Piper, Leonard, or Eileen J. Garrett, among others, should realize that such observations do not apply to them, or to many other mediums. Joseph Maxwell (1903/1905), for example, criticized Janet along these lines: "Up to the present Janet seems to have operated with invalids only, and I am not surprised, therefore, that he should assimilate the automatic phenomena of sensitives with those of his hysterical patients" (p. 261). He wrote further: "Hysterical people do not always give clear, decided phenomena; my best experiments have been made with those who were not in any way hysterical" (Maxwell, 1903/1905:44).

Others were also skeptical of pathology in mediums. In fact, one physician referred to such ideas as "false conceptions and legends" coming from observations of "hospital hysterics" (De Sermyn, 1910:133). Similarly, Charles Richet (1922:50) rejected the characterization of mediums as morbid individuals. In his view, they did not show more problems than other individuals.[12]

A possible line of research is that following Maxwell's (1903/1905) suggestion that the nervous system of mediums is liable to many changes and fluctuations that do not necessarily become pathological: "It seems to me that a certain impressionability—or nervous instability—is a favourable condition for the effervescence of mediany. I use the term *nervous instability* for want of a better one, but I do not use it in an ill sense" (p. 44). By nervous instability Maxwell said he did not mean pathology such as that found in hysteria, neurasthenia, or other afflictions. As he wrote:

> It is a state of the nervous system such as appears in hypertension. A lively impressionability, a delicate susceptibility, a certain unequalnessness of temper, establish analogy between mediums and certain neurotic patients; but they are to be distinguished from the latter by the integrity of their sensibilities, of their reflex movements, and of their visual range. As a rule, they have a lively intelligence, are susceptible to attention, and do not lack energy; their artistic sentiments are relatively developed; they are confiding and unreserved with those who show them sympathy; are distrustful and irritable if not treated gently. They pass easily from sadness to joy, and experience an irresistible need of physical agitation: These two characteristics are just the ones which made me choose the expression of nervous instability. I say instability, I do not say want of equilibrium. Many mediums whom I have known have an extremely well-balanced mind, from a mental and nervous point of view. My impression is that their nervous system is even superior to that of the average. (Maxwell, 1903/1905:44–45)

While Maxwell's descriptions do not seem to me to characterize the behaviors of many mediums, the idea of a mediumistic labile nervous system deserves exploration. In fact, the concept is consistent with some contemporary ideas of schizotypy that postulate that some people are more creative or hallucinatory than others due to differences in the inhibitory mechanisms of the nervous system (McCreery, 1997). Their lack of inhibition is postulated to cause arousal—which in turn is affected by situational variables—leading to a variety of experiences such as hallucinations. While it is unlikely that this is the whole picture for mediumship, the idea deserves to be explored through psychophysiological experimentation. More specifically, and in relation to Brazilian Candomble mediums, Seligman (2004) has hypothesized that mediums have an inability to regulate physiological arousal.

The study of the relationship between mediumship and psychopathology needs more empirical attention. Examples of what could be done can be found in recent research that does not support a relationship between pathology and mediumship (some recent research includes Moreira-Almeida, Lotufo Neto, & Greyson, 2007, Moreira-Almeida, Lotufo Neto, & Cardeña, 2008).

Further efforts could focus on old discussions of the similarities between alters seen in dissociative identity disorder and spirit controls (Hyslop, 1917: 12–29, Troubridge, 1922; see also Braude, 1988). Also, the topic may be explored following Janet's (1889) ideas that pathology may be related to mediumship through a common predisposition shared by both, and not necessarily as a simple cause and effect.[13] Similar to other research areas about relationships of psychological processes or phenomena to psychopathology (e.g., Barrantes-Vidal, 2003, Pickles & Hill, 2006), this hypothetical shared predisposition may have different pathways affected by both situational and developmental variables that may lead into adjustment or maladjustment.

Regardless of comparisons between mediums and non-mediums, there could be explorations about the possible existence of subgroups of mediums who show pathological tendencies. That is, while mediumship may not be generally related to psychopathology, there could be exceptions to the rule. This is perhaps similar to research conducted with presumably normal individuals who exhibit fantasy proneness, which is not in and of itself pathological, but may nonetheless be positively related to results on measures of pathology (Rhue & Lynn, 1987). Furthermore, we might also try to find evidence for Myers' (1903) view that the normal may use similar pathways of expression as the abnormal.[14]

Flournoy (1900) reminded us that Hélène Smith showed "disturbances of motility and sensibility, from which she seems entirely free in her normal state" (p. 441). While not all mediums have shown such disturbances, and I doubt many current mediums would, this brings up the study of neuropsychological dysfunctions. Possible research topics may include, among others, problems with orientation, attention, perception, memory, language skills, reasoning, and motor performance (Lezak, Horwieson, Loring, Hannay, & Fischer, 2004).

Another research possibility is that pointed out by such authors as Lombroso

(1909) and Morselli (1908) regarding physical mediumship. In their view, there could be individuals who exhibited simultaneous pathology and veridical phenomena. The idea is also consistent with the writings of psychotherapists such as Ehrenwald (1948).

Psychological and Parapsychological Experiences

From the days of William B. Carpenter (1853), to later times (Flournoy, 1900, Lebiedzinski, 1924, Sudre, 1946), the medium has been seen as a person with great imaginal potential.[15] However, there has been little research to put such assumptions to the test.

Much could be done today focusing on the constructs of absorption, boundary thinness, dissociation, fantasy proneness, hypnotic susceptibility, and transliminality, among others. Such constructs have shown positive relationships with parapsychological experiences in various questionnaire studies, suggesting the existence of a predisposition or a trait associated with these experiences that may be related to mediumship (Glicksohn, 1990, Hartmann, 1991, Myers, Austrin, Grisso, & Nickeson, 1983, Pekala, Kumar, & Marcano, 1995, Richards, 1991; see also discussion by Cardeña & Terhune, 2008).

The "disaggregation of personal perception" was described by Janet (1889:413) as a characteristic of both hypnosis and mediumship. In fact, dissociative experiences separate from mediumistic performances have been explored in a few modern studies with mediums (e.g., Hughes, 1992, Laria, 1998, Moreira-Almeida, Lotufo Neto, & Greyson, 2007, Negro, Palladino-Negro, Rodrigues Louzã, 2002, Reinsel, 2003, Roxburgh, 2008). While there is no doubt that mediums have dissociative experiences, as measured by standardized instruments, more could be done about the specific dissociative experiences they have and how they compare to the dissociative experiences of other individuals. One possible line of exploration, among many, is the relationship of mediumship to different types of dissociation (on the latter, see Brown, 2006, Cardeña, 1994). In addition to measures of absorption, depersonalization, and amnesia, this may also include somatoform dissociation (Nijenhuis, 2004).

Roll (1982) has suggested that some mediums have experiences suggestive of fantasy-proneness. These experiences include visions in childhood and difficulties in distinguishing fantasy from reality.

Some of the above-mentioned constructs—boundary thinness and transliminality—are theoretically related to ideas concerning various degrees of communication between the conscious and the subconscious mind. Such a notion was central to Myers' (1903) conception of such phenomena as hypnosis, creativity, hallucinations, and mediumship. According to Flournoy (1901): "All the difference between mediums and ordinary people is that in the latter there is a very pronounced gap . . . between wakefulness and daydreams. It is the opposite in mediums. . . ." (p. 127). Consistent with this idea, De Sermyn (1910) believed one of the defining features of mediums was "the facility with which the rapport between their conscious and unconscious will is produced" (p. 133). Scales to measure boundary thinness (Hartmann, 1991) and transliminality (Lange, Thalbourne, Houran, & Storm, 2000) may be employed in new research to test these ideas. Roxburgh (2008) did not find significant differences between mediums and non-mediums on boundary thinness. Furthermore, it is possible to conduct perceptual tests to measure the subliminal thresholds of mediums so as to assess directly if they have more perceptions of subliminal stimuli, perhaps indicating a more permeable barrier between conscious and subconscious than controls.

There is also a need to investigate other experiences mediums may have outside of their mediumistic performances. Charles Emmons' examination of autobiographies of mediums, and interviews with mediums, showed that their lives present many spontaneous parapsychological experiences (Emmons & Emmons, 2003). In their autobiographies, Gladys Osborne Leonard (1931) and Eileen J. Garrett (1939) mention that they had a variety of phenomena in their daily lives such as out-of-body experiences and apparitions. Mrs. Thompson saw spirits, heard voices, and perceived images in crystal balls (Myers, 1902:70). Hélène Smith had many experiences outside the seance room. In Flournoy's words:

> Hélène's spontaneous automatisms have often aided her in, without ever having interfered with, her daily occupations. There is, happily for her, a great difference in intensity between the phenomena of her seances and those which break in upon her habitual existence, the latter never having caused such disturbance of her personality as the former.
>
> In her daily life she has only passing hallucinations limited to one or two of the senses, superficial hemisomnambulisms, compatible with a certain amount of self-possession—in short, ephemeral perturbations of no importance from a practical point of view. Taken as a whole, the interventions of the subliminal in her ordinary existence are more beneficial to her than otherwise, since they often bear the stamp of utility and appropriateness, which make them very serviceable.
>
> Phenomena of hypermnesia, divination, lost objects mysteriously recovered, happy inspirations, true presentiments, correct intuitions . . . she possesses in so high a degree that this small coin of genius is more than sufficient to compensate for the inconveniences resulting from the distraction and momentary absence of mind with which the vision is accompanied. (Flournoy, 1900:47)

New studies eliciting information from mediums could compare both the prevalence, frequency, and variety of phenomena in their daily lives (e.g., hypnagogic imagery, ESP, and apparitional experiences), as assessed against control groups. In addition to expecting more experiences with mediums due to their presumed openness to a variety of influences, a possible prediction is that mediums would have more phenomena suggestive of discarnate agency (e.g., visual and auditory perceptions about the dead) than non-mediums. Furthermore, perhaps experienced mediums will show both a higher frequency and a higher level of control of these experiences than less-experienced mediums, and than non-mediums.

Another neglected area is the study of possible personality variables related to mediumship. Regarding the first, Schmeidler (1958) reported analyses of Rorschach responses generated by medium Caroline Chapman. The medium was said to show "little need for intimate personal involvement with others" (Schmeidler, 1958:153), as well as "inner meaning of events, symbolic values, [and] the appreciation of nature . . . ." (p. 153). A few years later, Trick (1966) reported finding a tendency for field dependence in two mediums. Roxburgh (2008) found positive associations between mediumship and extraversion, neuroticism, and openness to experience, but no significant associations with agreeableness or conscientiousness.[16]

Other possibilities for future research may come from the statement that mediums show "excessive emotionalism" (Morselli, 1908(1):97), and are mainly characterized by "their tendency to distraction, to daydreams . . ." (De Sermyn, 1910:133). Work on these topics could be done using measures of emotions (Plutchik, 1989), daydreaming (Singer & McCraven, 1961), and distraction (Broadbent, Cooper, FitzGerald, & Parkes, 1982).

Finally, we may focus on cognitive studies of mediums. This may include tests for memory inhibition, facilitation, and cognitive interference, such as those conducted with DID patients (Dorahy & Irwin, 2004). Other performance tests may include those used to assess verbal and spatial ability, and problem-solving, among others.

Does Mediumship Change over Time?

There are many observations suggesting that mediumship is a developmental process. I am not concerned here with methods to develop mediumship but instead with observations suggesting changes in mediumship over time. Examples include the following observations made by John W. Edmonds of changes in a medium that, unfortunately, are not precise about the lapse of time between the changes shown by the medium, described as follows:

> At first she was violently agitated in her person. She soon wrote mechanically; that is without any volition on her part, and without any consciousness

of what she was penning. . . . She next became a speaking medium. She was not entranced . . . but was fully aware of all she was saying and of all that occurred around her. She . . . was shown, through the instrumentality of her own mind, all the particulars of the wreck of the steamer *San Francisco*. . . . All this was several days before any news had reached the land of the accident to that vessel. . . . A few days brought minute confirmation of every incident which had been disclosed to her. . . .

She next became developed to speak different languages. She knows no language but her own, and a little smattering of boarding-school French. Yet she has spoken in nine or ten different tongues, sometimes for an hour at a time, with the ease and fluency of a native. . . .

About the same time her musical powers became developed. She has repeatedly sung in foreign languages, such as Italian, Indian, German, and Polish, and it is now not unfrequent that she sings in her own language, improvising both words and tune as she proceeds—the melody being very unique and perfect, and the sentiments in the highest degree elevating and ennobling.

Her next advance was to see spirits and spiritual scenes, and now scarcely a day passes that she does not describe the spirits who are present, entire strangers to her, yet very readily recognized and identified by their inquiring friends. . . .

At one time she was used as the instrument for delivering long and didactic discourses on the principles of our faith. Now she is mostly used to give moral and mental tests, which to many are very satisfactory. At one time she saw chiefly allegorical pictures; now she sees the reality of spiritual life. Once she wrote mechanically, but now by impression, knowing the thoughts she pens. Formerly it was difficult for spirits to converse through her; but now conversation, with anyone, however much a stranger to her, goes on with a freedom and ease most gratifying to the investigator. (Edmonds & Dexter, 1855(2):44–45)

The mediumship of Mrs. Piper showed changes over time both in the variety of spirit controls as well as in mode of expression. For example, the Phinuit control which was active from the 1880s to 1892 relied mainly on Mrs. Piper's voice to communicate while the G. P. control (from 1892 to 1897) relied more on writing (Sidgwick, 1915). In the case of Hélène Smith, the medium produced different phenomena in different time periods. After Flournoy published his famous study (Flournoy, 1900), a period during which he did not have access to the medium anymore, she went on to produce further Martian communications, and added new variants related to Uranus and to the Moon (Flournoy, 1901). Later on she presented what Lemaitre (1908) called a "new somnambulistic cycle" consisting of religious paintings. While Smith had produced paintings before, such as those related to Mars (Flournoy, 1900), she started having visions of Christ in 1900, and the first religious paintings appeared in 1903 and continued for several years.[17]

Evidence that mediums have had psychic experiences in childhood

(Emmons & Emmons, 2003, Chapter 52) suggests they may have had a predisposition to do so from the beginning. But there is a need for studies of the multiple factors—such as parenting, education, support, and training—that may have interacted with such a hypothetical predisposition across the medium's lifespan. Such work may allow us to identify common developmental pathways of mediumship.

Concluding Remarks

The purpose of this paper has been to suggest lines for future research with mental mediums using mainly the old psychical research literature as inspiration for ideas, with emphasis on psychological approaches. But such a complex phenomenon as mediumship needs to be studied in multiple ways, such as medically and anthropologically, among other perspectives that I have not emphasized in this paper (for a discussion of the psychosocial functions of mediumship, see Maraldi, 2009).

These studies encounter many problems. Among them is the difficulty of making sense of mediumship through relationships with other variables and processes that are not fully understood, such as dissociation, and the way the mind handles imagery. The lack of more detailed knowledge about the processes that complicate mentation in general—association of ideas, confusion of images, decoding of symbols—also hinders our progress in the understanding of mediumship.

Furthermore, our understanding of these manifestations may be complicated by the possibility that we may be dealing with different types of mediumship. While some cases of mediumship may be solely explained by the abilities of the medium to produce mentation based on psychological processes and social influences, other cases such as those of Piper and Leonard have presented veridical features and seem to require explanations beyond the conventional ones. Assuming these are different types of mediums, I wonder if they would differ in their relationship to the variables discussed in this paper.[18] This may not be the case if we postulate, following Myers (1903), that veridical and non-veridical mediumship share the same subconscious processes of use of sensory and motor automatisms to convey messages. Consequently, if many of the variables discussed in this paper—dissociation, hypnotic susceptibility, or an openness to all kinds of experiences—manifests through the subconscious mind of the medium, they would be related to mediumship whether they can produce veridical content or not.

Then there is the issue of the interaction of many variables. Perhaps a model may be developed considering the interaction of early childhood experiences, later situational influences, including mediumistic training, and personality and cognitive variables.[19]

While I believe these variables are important, it is also important to assess cultural differences. For example, how does mediumship in countries with different cultural backgrounds such as Brazil and the United States compare regarding the variables discussed in this paper? After all, the social and religious factors that shape mediumship in Brazil—Afro-Brazilian cults and Kardecist spiritism—are generally different from those in which most mediumship develops in the States.

There are of course no simple solutions. In addition to continuing research in the above-mentioned areas, and in others not covered here, we may make significant advances if we could design studies of mediums that collected information in the same project from a variety of perspectives. A model in point was a report published about Italian medium Eusapia Palladino in which she was studied from the physical, physiological, psychological, and parapsychological perspectives (Courtier, 1908). Perhaps we could conduct research with a mental medium today in which there would be psychophysiological monitoring during the performance, analyses of possible patterns in the mentation (e.g., recurrent and symbolic images, predominant sensory modalities, distortions), assessment of personality and cognitive variables, and veridical mentation, and collection of information about the medium's childhood, family, social environment, mediumistic training, psychic experiences, health, and possible changes over time in the phenomena. While such a research program may not be practical or possible from the financial point of view, in theory it could provide much useful information about various aspects of mediumship.

Other literatures about investigations of particular individuals that may inspire our studies of mediums include those conducted with individuals with dissociative identity disorder (e.g., Jeans, 1976, M. Prince, 1908), and with cases of spirit possession (Crapanzano & Garrison, 1977). Students of mediumship could also emulate studies of single individuals with exceptional abilities (Obler & Fein, 1988), such as prodigious memory (Luria, 1987, Parker, Cahill, & McGaugh, 2006), and studies of neuropsychological disorders (Ogden, 2005) and hallucinatory experiences (Harris & Gregory, 1981, Schatzman, 1980). Today one remembers with respect the detailed studies done with mediums Mrs. Willett (Balfour, 1935), Hélène Smith (Deonna, 1932, Flournoy, 1900, 1901), and Mrs. Piper (Sidgwick, 1915), which could be used to guide further research. New studies could also focus on particular phenomena, as seen in past work about specific mediumistic phenomena such as paintings (Deonna, 1932), literary productions (W. F. Prince, 1927), and trance personalities (Progoff, 1964).[20]

"The exact nature of mediumship," wrote Horace Leaf (1919:125), "promises long to remain a mystery." Nonetheless, the difficulties inherent in the study of the subject should not stop the research. It is my belief that the past literature, used in conjunction with modern ideas, methodology, and creativity, can help researchers focus their efforts to increase our understanding of mental mediumship.

Acknowledgements

I wish to thank Nancy L. Zingrone for useful editorial suggestions.

Notes

[1] There are also many interesting observations about trance in the literature of physical mediums (e.g., Lombroso, 1909, Morselli, 1908).

[2] Although the focus of this paper is not on psychophysiology, it is interesting to note that William James (1886:105) mentioned that Piper's pupils contracted during trance. Hodgson (1892:5) simply stated that Piper's pupils reacted to light in trance. It is to be hoped that future studies of these issues will both be more systematic and reported in more detail.

[3] Capron (1855:381) referred to the difficulties in separating spirit influence from that of the medium's mind. Kardec (1862, Chapter 19) believed that the medium could influence communications and that the content of his or her brain was used to form messages. Myers (1903(2):250) referred to a possible mix in trance communications of "mixed telepathy between the sitter, the sensitive's spirit, and the extraneous spirit." In his view, to separate possible influences one had to be familiar with the medium in question. This last statement reminds us of the importance of studying mediums over time so as to have information about the features of their mentation.

[4] Symbols are not necessarily limited to mentation, as seen in the production of symbolic drawings such as religious scenes and emblems described by Crosland (1857).

[5] Many writers have addressed the fragmentary nature of mediumistic communications in terms of both coherence and veridical content. Referring to Mrs. Piper, Oliver Lodge (1890) wrote: "In the midst of . . . lucidity a number of mistaken and confused statements are frequently made, having little or no apparent meaning or application" (p. 443).

[6] Following the previously mentioned writers, Myers (1900:402) suggested that Kardec conveyed suggestions to his mediums.

[7] James (1890), always fair to the opposite point of view, commented after the above quote as follows: "But the odd thing is that persons unexposed to spiritualist traditions will so often act in the same way when they become entranced, speak in the name of the departed, go through the motions of their several death-agonies, send messages about their happy home in the summer-land, and describe the ailments of those present" (James, 1890(1):394).

[8] Twenty years later, Sudre (1946) published a book discussing personation phenomena in mediumship and in other phenomena. Unfortunately, the book does not seem to be well-known today.

[9] These topics, and philosophical ones, have been discussed frequently by mediumistic communicators (e.g., Kelway-Bamber, 1920, Linton, 1855, Moses, 1883, *Thoughts*, 1886). Kardec (1862, 1867) developed a system out of communications which he used to provide a philosophical basis for spiritism and explanations for psychic phenomena. Bayless (1971) and Fontana (2009) discuss many examples of the "teachings" of the spirits received through mediums, with emphasis on the nature of the afterlife.

[10] Another example of veridical and non non-veridical communications coming from Leonard can be found in Oliver Lodge's (1916) well-known discussion of messages presumably from his son.
[11] The interpretation of this is complicated by Hall's admission that there was a living person by that name whom he may have known about (Tanner, 1910:181).
[12] Many others have objected to the pathologizing of mediums (e.g., Kardec, 1862:263, Sprague, 1912:44–45).
[13] Le Maléfan (1999) discusses different models of the postulated pathology–mediumship relationship, as seen in past French psychiatric literature.
[14] According to Myers (1903):

> *It may be expected that supernormal vital phenomena will manifest themselves as far as possible through the same channels as abnormal or morbid vital phenomena, when the same centres or the same synergies are involved*' (Myers, 1903(2):84, italics in the original).

He further wrote about manifestations of a secondary self that

> it seems probable that its readiest path of externalisation—its readiest outlet of visible action,—may often lie along some track which has already been shown to be a line of low resistance by the disintegrating processes of disease .... If epilepsy, madness, & c., tend to split up our faculties in certain ways, automatism is likely to split them up in ways somwhat [sic] resembling these. (Myers, 1903(2):84)

[15] I will discuss dreams and mediumship in another paper. Here I will only mention as a hypothesis that mediums, as compared to controls, may show a higher frequency of dream recall, and more reports of lucid, vivid, and unusual dreams.
[16] See also Hearne (1989). Ideas for research can also be found in the literature about psychological characteristics of psychics (e.g., Schmeidler, 1982, Tenhaeff, 1972). There is also the issue of creativity in mediums, a topic discussed by Braude (2000) and Grosso (1997), and researched by Roxburgh (2008) with null results. Flournoy's (1900, 1901) and Deonna's (1932) writings focus on Héléne Smith's creations. The latter is a much-neglected study of mediumistic paintings, a topic with implications for the mediumship–creativity relationship (see also Osty, 1928).
[17] The most detailed study of Smith's paintings, and a unique study in the literature of mediumship, was Waldemar Deonna's (1932). He argued that some of the reasons for changes in the phenomena were the medium's reactions to Flournoy's (1900) psychological analysis, and the fact that, due to a patron, she stopped working for a living and devoted her life only to her mediumship. This loss of her normal activities, Deonna speculated, led her subconscious mind to pay more attention to hallucinatory commands that "seemed to come from the beyond" (Deonna, 1932:59).
[18] In practice, it is not that simple to separate mediums in terms of their production of veridical and non-veridical mentation. As seen in the case of Mrs. Leonard (Lodge, 1916, Thomas, 1928), mediums who have produced veridical phenomena also produce non-veridical ones that may have as their source only the intrapsychic processes of the medium. The mediumistic process, as Myers (1903) and Hyslop (1919) stated, is probably a mix of different influences.
[19] On the interactions of some of these variables in Brazilian mediumship, see Krippner (1989) and Seligman (2005).
[20] Unfortunately, a good portion of modern psychological and psychiatric research has moved away from in-depth studies of specific individuals (for discussions, see Berkenkotter, 2008, Danziger, 1990).

References

Aksakof, A. (1875). Researches on the historical origin of the reincarnation speculations of French spiritualists. *Spiritualist Newspaper*, August 13, 1875, pp. 74–75.

Allan Kardec's "Spirits" book (1875). *Spiritualist Newspaper*, October 8, 1875, pp. 169–170.

Alvarado, C. S. (1991). Iatrogenesis and dissociation: A historical note. *Dissociation, 4,* 36–38.

Alvarado, C. S. (1998). ESP and altered states of consciousness: An overview of conceptual and research trends. *Journal of Parapsychology, 62,* 27–63.

Alvarado, C. S., Machado, F. R., Zangari, W., & Zingrone, N. L. (2007). Perspectivas históricas da influência da mediunidade na construção de idéias psicológicas e psiquiátricas. *Revista de Psiquiatria Clínica, 34*(Supp.1), 42–53.

Bailey, D. E. (1886). *Thoughts from the Inner Life.* Boston: Colby & Rich.

Balfour, G. W., Earl of (1935). A study of the psychological aspects of Mrs. Willett's mediumship, and of the statements of the communicators concerning process. *Proceedings of the Society for Psychical Research, 43,* 41–318.

Ballet, G. (1913). La psychose hallucinatoire chronique et la désagrégation de la personnalité. *L'encéphale, 8,* 501–519.

Barkas, T. P. (1862). *Outlines of Ten Years' Investigations into the Phenomena of Modern Spiritualism, Embracing Letters, & c.* London: Frederick Pitman.

Barrantes-Vidal, N. (2003). Creativity and madness revisited from current psychological perspectives. *Journal of Consciousness Studies, 11,* 58–78.

Barrett, K. W. (1996). *A Phenomenological Study of Channeling: The Experience of Transmitting Information from a Source Perceived as Paranormal.* [Ph.D. dissertation]. Palo Alto, CA: Institute of Transpersonal Psychology.

Bayless, R. (1971). *The Other Side of Death.* New Hyde Park, NY: University Books.

Beischel, J., & Schwartz, G. E. (2007). Anomalous information reception by research mediums demonstrated using a novel triple-blind protocol. *Explore: The Journal of Science & Healing, 3,* 23–27.

Berkenkotter, C. (2008). *Patient Tales: Case Histories and the Uses of Narratives in Psychiatry.* Columbia, SC: University of South Carolina Press.

Binet, A. (1892). *Les Altérations de la Personnalité.* Paris: Félix Alcan.

Bozzano, E. (1907). Symbolism and metapsychical phenomena. *Annals of Psychical Science, 6,* 235–259.

Braude, S. E. (1988). Mediumship and multiple personality. *Journal of the Society for Psychical Research, 55,* 177–195.

Braude, S. E. (2000). Dissociation and latent abilities: The strange case of Patience Worth. *Journal of Trauma and Dissociation, 1,* 13–48.

Broadbent, D. E., Cooper, P. F., FitzGerald, P., & Parkes, K. R. (1982). The Cognitive Failures Questionnaire (CFQ) and its correlates. *British Journal of Clinical Psychology, 21,* 1–16.

Brown, R. J. (2006). Different types of 'dissociation' have different psychological mechanisms. *Journal of Trauma and Dissociation, 7,* 7–28.

Capron, E. W. (1855). *Modern Spiritualism: Its Facts and Fanaticisms, Its Consistencies and Contradictions.* Boston: Bela Marsh.

Cardeña, E. (1994). The domain of dissociation. In: S. J. Lynn and J. W. Rhue (Eds.). *Dissociation: Clinical, Theoretical, and Research Perspectives* (pp. 15–31). New York: Guilford.

Cardeña, E., & Terhune, D. (2008). A distinct personality trait? The relationship between hypnotizability, absorption, self-transcendence, and mental boundaries. Paper presented at the 51st Annual Conference of the Parapsychological Association; University of Winchester, UK; August 13–17, 2008.

Cardeña, E., Terhune, D. B., Lööf, A., & Buratti, S. (2009). Hypnotic experience is related to emotional contagion. *International Journal of Clinical and Experimental Hypnosis, 57,* 33–46.

Carpenter, W. B. (1853). Electrobiology and mesmerism. *Quarterly Review, 93,* 501–557.

Carrington, H. (1920). *Your Psychic Powers and How To Develop Them.* New York: Dodd, Mead.

Courtier, J. (1908). Rapport sur les séances d'Eusapia Paladino à l'Institut Général Psychologique. *Bulletin de l'Institut Général Psychologique, 8,* 415–546.

Crapanzano, V., & Garrison, V. (1977). *Case Studies in Spirit Possession.* New York: Wiley.
Crosland, Mrs. N. (1857). *Light in the Valley: My Experiences of Spiritualism.* London: G. Routledge.
Danzinger, K. (1990). *Constructing the Subject: Historical Origins of Psychological Research.* Cambridge: Cambridge University Press.
Delboeuf, J. R. L. (1886). De l'Influence de l'Éducation et de l'Imitation dans le Somnambulisme Provoqué. *Revue Philosophique de la France et de l'Étranger, 22,* 146–171.
The delusion of spiritualism (1860). *Lancet, 2,* 466–467.
Deonna, W. (1932). *De la Planète Mars en Terre Sainte: Art et Subconscient.* Paris: E. de Boccard.
Dorahy, M. J., & Irwin, H. J. (2004). Assessing markers of working memory function in dissociative identity disorder using neutral stimuli: A comparison with clinical and general population samples. *Australian and New Zealand Journal of Psychiatry, 38,* 47–55.
Edmonds, J. W., & Dexter, G. T. (1855). *Spiritualism* (Vol. 2). New York: Partridge & Brittan.
Ehrenwald, J. (1948). *Telepathy and Medical Psychology.* New York: W. W. Norton.
Emmons, C. F., & Emmons, P. (2003). *Guided by Spirit: A Journey into the Mind of the Medium.* New York: Writers Club Press.
Evans, C. C., & Osborn, E. (1952). An experiment in the electro-encephalography of mediumistic trance. *Journal of the Society for Psychical Research, 36,* 588–596.
Flammarion, C. (1907). *Mysterious Psychic Forces: An Account of the Author's Investigations in Psychical Research, Together with Those of Other European Savants.* Boston: Small, Maynard.
Flournoy, T. (1900). *From India to the Planet Mars: A Study of a Case of Somnabulism.* New York: Harper & Brothers.
Flournoy, T. (1901). Nouvelles observations sur un cas de somnambulisme avec glossolalie. *Archives de Psychologie, 1,* 101–255.
Flournoy, T. (1911). *Spiritism and Psychology.* New York: Harper & Brothers.
Fontana, D. (2009). *Life Beyond Death: What Should We Expect?* London: Watkins.
Frigerio, A. (1989). Levels of possession awareness in Afro–Brazilian religions. *AASC Quarterly, 5*(2–3), 5–11.
Garrett, E. J. (1939). *My Life as a Search for the Meaning of Mediumship.* New York: Oquaga Press.
Glicksohn, J. (1990). Belief in the paranormal and subjective paranormal experience. *Personality and Individual Differences, 11,* 675–683.
Grosso, M. (1997). Inspiration, mediumship, surrealism: The concept of creative dissociation. In: S. Krippner and S. M. Powers (Eds.), *Broken Images, Broken Selves: Dissociative Narratives in Clinical Practice* (pp. 181–198). Philadelphia: Brunner/Mazel.
Harriman, P. L. (1942). The experimental production of some phenomena related to the multiple personality. *Journal of Abnormal and Social Psychology, 37,* 244–255.
Harris, J., & Gregory, R. L. (1981). Tests of the hallucinations of 'Ruth.' *Perception, 10,* 351–354.
Hartmann, E. (1991). *Boundaries of the Mind: A New Psychology of Personality.* New York: Basic Books.
Hearne, K. M. T. (1989). A questionnaire and personality study of self-styled psychics and mediums. *Journal of the Society for Psychical Research, 55,* 404–411.
Hodgson, R. (1892). A record of observations of certain phenomena of trance. *Proceedings of the Society for Psychical Research, 8,* 1–167.
Hodgson, R. (1898). A further record of observations of certain phenomena of trance. *Proceedings of the Society for Psychical Research, 13,* 284–582.
Hughes, D. J. (1992). Differences between trance channeling and multiple personality disorder on structured interview. *Journal of Transpersonal Psychology, 24,* 181–192.
Hyslop, J. H. (1906). The Smead case. *Annals of Psychical Science, 4,* 69–105.
Hyslop, J. H. (1917). The Doris case of multiple personality: Part III. *Proceedings of the American Society for Psychical Research, 11,* 5–866.
Hyslop, J. H. (1918). *Life After Death: Problems of the Future Life and Its Nature.* New York: E. P. Dutton.
Hyslop, J. H. (1919). *Contact with the Other World.* New York: Century.
Inglis, B. (1984). *Science and Parascience: A History of the Paranormal, 1914–1939.* London: Hodder and Stoughton.
James, W. (1886). Report of the Committee on Mediumistic Phenomena. *Proceedings of the*

American Society for Psychical Research, 1, 102–106.
James, W. (1890). *The Principles of Psychology* (2 vols.). New York: Henry Holt.
Janet, P. (1889). *L'Automatisme Psychologique: Essai de Psychologie Expérimentale sur les Formes Inférieures de l'Activité Humaine.* Paris: Félix Alcan.
Janet, P. (1909). Dèlire systématique à la suite de pratiques du spiritisme. *L'Encéphale, 4,* 363–367.
Jeans, R. F. (1976). The three faces of Evelyn: A case report: I. An independently validated case of multiple personality. *Journal of Abnormal Psychology, 85,* 249–255.
Jung, C. G. (1983). On the psychology and pathology of so-called occult phenomena. In: C. G. Jung, *Psychiatric Studies* (pp. 3–88). Princeton: Princeton University Press. (Original work published, 1902)
Kardec, A. (1862). *Spiritisme Expérimental: Le Livre des Médiums* (2nd ed.). Paris: Didier.
Kardec, A. (1867). *Philosophie Spiritualiste: Le Livre des Esprits* (15th ed.). Paris: Didier.
Kelway-Bamber, L. (Ed.). (1920). *Claude's Second Book.* New York: Henry Holt.
Krippner, S. (1989). A call to heal: Entry patterns in Brazilian mediumship. In: C. A. Ward (Ed.), *Altered States of Consciousness and Mental Health: A Cross-Cultural Perspective* (pp. 186–206). Thousand Oaks, CA: Sage.
Lange, R., Thalbourne, M.A., Houran, J., & Storm, L. (2000). The revised Transliminality Scale: Reliability and validity data from a Rasch top-down purification procedure. *Consciousness and Cognition, 9,* 591–617.
Laria, A. J. (1998). *Dissociative Experiences among Cuban Mental Health Patients and Spiritist Mediums.* [Ph.D. dissertation]. University of Massachusetts Boston.
Leaf, H. (1919). *What Is This Spiritualism?* New York: George H. Doran.
Leaf, W. (1890). A record of observations of certain phenomena of trance (3). Part II. *Proceedings of the Society for Psychical Research, 6,* 558–646.
Lebiedzinski, P. (1924). L'idéoplastic comme hypothèse directrice des études métapsychiques. *L'etat Actuel des Recherches Psychiques d'après les Travaux du IIme Congrès International Tenu à Varsovie en 1923 en l'Honneur du Dr Julien Ochorowicz* (pp. 286–300). Paris: Presses Universitaires de France.
Lemaitre, A. (1908). Un nouveau cycle somnambulique de Mlle Smith: Ses peintures religeuses. *Archives de Psychologie, 7,* 63–83.
Le Maléfan, P. (1999). *Folie et Spiritisme: Histoire du Discourse Psychopathologique sur la Pratique du Spiritisme, Ses Abords et Ses Avatars (1850–1950).* Paris: L'Hartmattan.
Leonard, G. O. (1931). *My Life in Two Worlds.* London: Cassell.
Levy, D.A., & Nail, P. R. (1993). Contagion: A theoretical and empirical review and reconceptualization. *Genetic, Social, and General Psychology Monographs, 119,* 233–284.
Lévy-Valensi, J. (1910). Spiritisme et folie. *L'Encéphale, 5,* 696–716.
Lezak, M. D., Horwieson, H. J., Loring, D. W., with Hannay, H. J., & Fischer, J. S. (2004). *Neuropsychological Assessment* (4th ed.). New York: Oxford University Press.
Linton, C. (1855). *The Healing of the Nations* (2nd ed.). New York: Society for the Diffusion of Spiritual Knowledge.
Lodge, O. (1890). A record of observations of certain phenomena of trance (2). Part I. *Proceedings of the Society for Psychical Research, 6,* 443–557.
Lodge, O. J. (1916). *Raymond, or Life after Death.* New York: George H. Doran.
Lombroso, C. (1909). *After Death—What? Spiritistic Phenomena and Their Interpretation.* Boston: Small, Maynard.
Luria A. R. (1987). *The Mind of a Mnemonist.* Cambridge, MA: Harvard University Press.
Mammarella, I. C., Pazzaglia, F., & Cornoldi, C. (2006). The assessment of imagery and visuo–spatial working memory functions in children and adults. In: T. Vecchi & G. Bottini (Eds.), *Imagery and Spatial Cognition: Methods, Models, and Cognitive Assessment* (pp. 15–38). Amsterdam: Benjamins.
Maraldi, E. de O. (2009). Un estudio exploratorio sobre la mediumnidad y la identidad psicosocial. *E-Boletin Psi, 4*(2). http://www.alipsi.com.ar/e-boletin/e_boletin_psi_4-2_Mayo_2009.htm#arriba02
Maxwell, J. (1905). *Metapsychical Phenomena: Methods and Observations.* London: Duckworth. (Original work published 1903)
McCreery, C. (1997). Hallucinations and arousability: Pointers to a theory of psychosis. In: G. Claridge (Ed.), *Schizotypy: Implications for Illness and Health* (pp. 251–273). Oxford:

Oxford University Press.
Moreira-Almeida A., Almeida, A. A. S., & Lotufo Neto, F. (2005). History of spiritist madness in Brazil. *History of Psychiatry, 16,* 5–25.
Moreira-Almeida A., Lotufo Neto, F., & Cardeña E. (2008). Comparison of Brazilian spiritist mediumship and dissociative identity disorder. *Journal of Nervous and Mental Disease, 196,* 420–424.
Moreira-Almeida A., Lotufo Neto, F., & Greyson, B. (2007). Dissociative and psychotic experiences in Brazilian spiritist nediums. *Psychotherapy and Psychosomatics, 76,* 57–58.
Morselli, E. (1908). *Psicologia e "Spiritismo:" Impressioni e Note Critiche sui Fenomeni Medianici di Eusapia Paladino* (2 vols.). Turin: Fratelli Bocca.
Moses, W. S. [under pseudonym M. A. Oxon] (1883). *Spirit Teachings.* London: Psychological Press Association.
Myers, F. W. H. (1884). On a telepathic explanation of some so-called spiritualistic phenomena: Part I. *Proceedings of the Society for Psychical Research, 2,* 217–237.
Myers, F. W. H. (1885). Automatic writing. II. *Proceedings of the Society for Psychical Research, 3,* 1–63.
Myers, F. W. H. (1900). Pseudo-possession. *Proceedings of the Society for Psychical Research, 15,* 384–415.
Myers, F. W. H. (1902). On the trance-phenomena of Mrs. Thompson. *Proceedings of the Society for Psychical Research, 17,* 67–74.
Myers, F. W. H. (1903). *Human Personality and Its Survival of Bodily Death* (2 vols.). London: Longmans, Green.
Myers, S. A., Austrin, H. R., Grisso, J. T., & Nickeson, R. C. (1983). Personality characteristics as related to the out-of-body experience. *Journal of Parapsychology, 47,* 131–144.
Negro, P. J., Palladino-Negro, P., & Rodrigues Louza, M. (2002). Do religious mediumship dissociative experiences conform to the sociocognitive theory of dissociation? *Journal of Trauma and Dissociation, 3,* 51–73.
Nijenhuis, E. R. S. (2004). *Somatoform Dissociation: Phenomena, Measurement, and Theoretical Issues.* New York: W. W. Norton.
Obler, L. K., & Fein, D. (1988). *The Exceptional Brain: Neuropsychology of Talent and Special Abilities.* New York: Guilford Press.
Ogden, J. A. (2005). *Fractured Minds: A Case-Study Approach to Clinical Neuropsychology* (2nd ed.). New York: Oxford University Press.
Osty, E. (1928). Aux confins de la psychologie classique et de la psychologie métapsychique: II. M. Augustine Lesage, peintre sans avois apris. *Revue métapsychique, 1,* 1–35.
Parker, E. S., Cahill, L., & McGaugh, J. L. (2006). A case of unusual autobiographical remembering. *Neurocase, 12,* 35–49.
Pekala, R. J. (1991). *Quantifying Consciousness: An Empirical Approach.* New York: Plenum Press.
Pekala, R. J., & Kumar, V. K. (2000). Operationalizing 'trance.' I: Rationale and research using a psychophenomenological approach. *American Journal of Clinical Hypnosis, 43,* 107–135.
Pekala, R. J., Kumar, V. K., & Marcano, G. (1995). Anomalous/paranormal experiences, hypnotic susceptibility, and dissociation. *Journal of the American Society for Psychical Research, 89,* 313–332.
Pickles, A., & Hill, J. (2006). Developmental pathways. In: D. Cicchetti & D. J. Cohen (Eds.), *Developmental Psychology* (2nd ed., Vol. 1, pp. 211–243). Hoboken, NJ: John Wiley.
Plutchik, R. (Ed.). (1989). *The Measurement of Emotions.* San Diego, CA: Academic Press.
Prince, M. (1908). *The Dissociation of a Personality* (2nd ed.). London: Longmans, Green.
Prince, W. F. (1927). *The Case of Patience Worth: A Critical Study of Certain Unusual Phenomena.* Boston: Boston Society for Psychic Research.
Progoff, I. (1964). *The Image of an Oracle: A Report on Research into the Mediumship of Eileen J. Garrett.* New York: Garrett Publications.
Radclyffe-Hall, M., & Troubridge, U. (1919). On a series of sittings with Mrs. Osborne Leonard. *Proceedings of the Society for Psychical Research, 30,* 339–554.
Reinsel, R. (2003). Dissociation and mental health in mediums and sensitives: A pilot survey. In: S. Wilson (Ed.), *Proceedings of the Parapsychological Association* (pp. 200–221).

Parapsychological Association.
Rhue, J. W., & Lynn, S. J. (1987). Fantasy proneness and psychopathology. *Journal of Personality and Social Psychology, 53,* 327–336.
Richards, D. G. (1991). A study of the correlation between subjective psychic experiences and dissociative experiences. *Dissociation, 4,* 83–91.
Richardson, J. T. E. (2006). Early methods for assessing imagery and nonverbal abilities. In: T. Vecchi & G. Bottini (Eds.), *Imagery and Spatial Cognition: Methods, Models and Cognitive assessment* (pp. 3–14). Amsterdam: Benjamins.
Richet, C. (1883). La personnalité et la memoire dans le somnambulisme. *Revue philosophique de la France et de l'Étranger, 15,* 225–242.
Richet, C. (1922). *Traité de Métapsychique.* Paris: Félix Alcan.
Riding, R. J. (2006). Cognitive style: A review. In: R. J. Riding & S. Rayner (Eds.), *Cognitive Styles* (pp. 315–344). Stamford, CT: Ablex.
Rock, A. J., Beischel, J., & Schwartz, G. E. (2008). Thematic analysis of research medium's experiences of discarnate communication. *Journal of Scientific Exploration, 22,* 179–192.
Rogers, E. C. (1853). *Philosophy of Mysterious Agents, Human and Mundane; or, the Dynamic Laws and Relations of Man.* Boston: J. P. Jewett.
Roll, W. G. (1982). Mediums and RSPK agents as fantasy-prone individuals. In: W. G. Roll, R. L. Morris, & R. A. White (Eds.), *Research in Parapsychology 1981* (pp. 42–44). Metuchen, NJ: Scarecrow Press.
Rouget, G. (1985). *Music and Trance: A Theory of the Relation between Music and Possession.* Chicago: University of Chicago Press. (First published in French 1980)
Roxburgh, E. (2008). The psychology of spiritualist mental mediumship. Paper presented at the 51st Annual Convention of the Parapsychological Association; University of Winchester, UK; August 13–17, 2008.
Saltmarsh, H. F. (1929). Report on the investigation of some sittings with Mrs. Warren Elliott. *Proceedings of the Society for Psychical Research, 39,* 47–184.
Sardou, V. (1858). Des habitations de la planète Jupiter. *Revue Spirite, 1,* 223–232.
Schatzman, M. (1980). *The Story of Ruth.* New York: Putnam.
Schmeidler, G. R. (1958). Analysis and evaluation of proxy sittings with Mrs. Caroline Chapman. *Journal of Parapsychology, 22,* 137–155.
Schmeidler, G. R. (1982). A possible commonality among gifted psychics. *Journal of the American Society for Psychical Research, 76,* 53–58.
Schwartz, B. L. (2002). *Tip-of-the-Tongue States: Phenomenology, Mechanism, and Lexical Retrieval.* Mahwah, NJ: Lawrence Erlbaum.
Seligman, R. A. (2004). *Sometimes Affliction Is a Door: A Bio–Psycho–Cultural Analysis of the Pathways to Candomble Mediumship.* [Ph.D. dissertation]. Emory University.
Seligman, R. A. (2005). Distress, dissociation, and embodied experience: Reconsidering the pathways to mediumship and mental health. *Ethos, 33,* 71–99.
De Sermyn, W. C. (1910). *Contribution à L'Étude de Certaines Facultés Cérébrales Méconnues.* Lausanne: Payot.
Sidgwick, Mrs. H. [E. M.]. (1915). A contribution to the study of the psychology of Mrs. Piper's trance phenomena. *Proceedings of the Society for Psychical Research, 28,* 1–657.
Singer, J. L., & McCraven, V. G. (1961). Some characteristics of adult dreaming. *Journal of Psychology: Interdisciplinary and Applied, 51,* 151–164.
Solfvin, G. F., Roll, W. G., & Kelly, E. G. (1977). A psychophysiological study of mediumistic communicators. *Parapsychology Review, 8*(3), 21–22.
Sprague, E. W. (1912). *Spirit Mediumship.* Detroit, MI: Author.
Stevens, R. (2000). Phenomenological approaches to the study of conscious awareness. In: M. Velmans (Ed.), *Investigating Phenomenal Consciousness: New Methodologies and Maps* (pp. 99–120). Amsterdam: J. Benjamins.
Sudre, R. (1926). *Introduction à la Métapsychique Humaine.* Paris: Payot.
Sudre, R. (1946). *Personnages d'au-Dela.* Paris: Société des Éditions de Noël.
Tanner, A. (1910). *Studies in Spiritism.* New York: D. Appleton.
Tenhaeff, W. H. C. (1972). *Telepathy and Clairvoyance: Views of Some Little Investigated Capabilities of Man.* Springfield, IL: Charles C. Thomas.

Thomas, C. D. (1928). *Life beyond Death, with Evidence*. London: W. Collins Sons.
Thomas, C. D. (1939). A proxy experiment of significant success. *Proceedings of the Society for Psychical Research, 45,* 257–306.
Trick, O. L. (1966). Psychological studies of two "mediums." *Journal of Parapsychology, 30,* 301–302. (Abstract)
Tromp, M. (2006). *Altered States: Sex, Nation, Drugs, and Self-Transformation in Victorian Spiritualism*. Albany, NY: State University of New York Press.
Troubridge, U. (1922). The modus operandi in so-called mediumistic trance. *Proceedings of the Society for Psychical Research, 32,* 344–278.
Von Hartmann, E. (1885). Spiritism. *Light, 5,* 405–409, 417–421, 429–432, 441–443, 453–456, 466–470, 479–482, 491–494.
Weiss, S. (1905). *Journeys to the Planet Mars or Our Mission to Ento (Mars)* (2nd ed.). Rochester, NY: Austin Publishing Company.
Wooffitt, R. (2006). *The Language of Mediums and Psychics: The Social Organization of Everyday Miracles*. Aldershot: Ashgate.

ESSAY

Advantages of Being Multiplex

MICHAEL GROSSO

Division of Perceptual Studies, University of Virginia, Charlottesville, Virginia 22902, USA
gross.michael@gmail.com; michaelgrosso.net

Abstract—This is the study of the creative potential of mediumship. It emphasizes a neglected and under-conceptualized form of creativity in the realm of personality development. Examples include the cases of mediums Pearl Curran, Hélène Smith, and Matthew Manning. Artists with mediumistic propensities are stressed, such as Keats, Rimbaud, Blake, Yeats, and James Merrill. There is discussion of Myers' theory of genius and how this bears on the concept of personal transformation. Myers' theory is used to shed light on Surrealism, outsider art, and the lives of Socrates and Joan of Arc. All this is comprehended under the Keatsian rubric of "soulmaking."

Keywords: mediumship—creativity—multiplicity

Introduction

If we are multiplex beings, let us gain the advantage of our multiplicity.
—Myers, 1886–1887:19

The term *mediumship* suggests communication with the dead, receiving impressions from another world, from gods and spirits. I will lay aside this question-begging sense of the term, and use terms such as *sensitive* or *automatist*. An action may be said to be automatic ("self-moved") when "it is determined in an organism apart from the central will or control of that organism" (Myers, 1896:168). The suggestion is that there are other centers of will and other "control" systems. This apparent experiential plasticity of the ego-sense, the agility with which the self under special circumstances may reconfigure its identity, is a psychological fact of great interest. This is the sense of *mediumship*—and its relationship to creativity—I want to focus on here.

Mediumistic performance may be associated with creativity in at least two senses: There are mediums who produce or perform "creative" works, and there are individuals noted for their literary or artistic gifts who reveal "mediumistic" or dissociated traits (e.g., in Ghiselin, 1952). My aim here is to clarify the relationship between mediumship and creativity in light of F. W. H. Myers' views of genius, automatism, and the subliminal mind (Myers, 1903). Myers' theory points to new domains of creative performance; for example, Surrealism, the art of the insane, and so-called "outsider" art. The broad question is how mediumistic, or dissociated, states increase the potential for personality development.

Definition

On the term *creativity,* following Myers let us say that a performance is creative if it produces something extraordinary and original in a particular domain, and if the product is subsequently widely agreed to be useful, or to embody some intrinsic human value such as beauty, truth, wisdom, and so forth.

Great names such as Shakespeare, Michelangelo, Mozart spring to mind but can be intimidating. Myers' theory of genius, while fully affirming the value of the giants and founders of culture, also points us toward numerous gradations and qualities of the creative process. Creativity can occur in the anonymous valley as well as on the peaks. Myers' theory puts us in touch with the whole of our psychic life and the range of creative performance proliferates. It is this expansion of the range of creative performance that I want to underscore, especially in the domain of personality development. There is a question, experimental and deeply philosophical, of applying one's genius to the art of what we may speak of as *self-creation.* May we not utilize our highest creative capacities as a means of heightening or reforming the human personality? This seems a neglected topic of scientific investigation, although it is anticipated in the science fiction of Robert Louis Stevenson's *Dr. Jekyll and Mr. Hyde* (Stevenson, 1966).

The Medium as Artist

Two types of experience show linkage between mediumship (or certain forms of automatism) and creativity. In the first, a medium or sensitive may "channel" a work of art, music, philosophy, or literature. It sometimes happens by chance that in playing with the ouija board, or attempting automatic writing, one discovers the ability to write poetry or draw or compose music.

One famous example is the American Pearl Curran, a Midwestern housewife of modest education. Mrs. Curran's life changed at age 31 when an experiment with a ouija board produced a personality calling itself Patience Worth, a self-declared 17th-century Englishwoman (Prince, 1964).

From 1913 to 1937, Pearl Curran produced via ouija board and speech automatisms 29 volumes of recorded communications from "Patience Worth." With spice and attitude, Patience poured out plays, novels, poems, witticisms, repartees, proverbs, a consistent virtuoso literary and linguistic performance. What is especially striking is that this new, creative personality seems to have sprung suddenly and completely into being from, or through, the relatively staid, apparently ordinary Pearl Curran.

Patience produced works in a variety of styles from 19th century Victorian to early Anglo-Saxon; no one quite knows how she was able to pack her novels with so many historically and linguistically apt items and references. Her novel *The Sorry Tale,* about the life and times of Jesus, was hailed by critics, and according to the *The New York Times* was "constructed with the precision and accuracy of a master hand."

The manner in which she produced her oeuvre was extraordinary: She rapidly improvised poems on any theme suggested. Her mnemonic feats rivaled those of so-called "savants": She could pick up on a composition broken off mid-sentence, and return to the same word and continue with the same fluency days or weeks later. When challenged to vary composing works in different styles, she complied. When Walter Prince asked Patience to dictate a poem to her husband John Curran *and* simultaneously write a letter to a friend, she passed the test.

None of this serves to prove survival of the dead. It does, however, enlarge our conception of what creative faculties may be latent in ordinary human personality. Given the evidence we have, the English pragmatist Schiller (1928) and more recently the American philosopher Braude (2003) describe "Patience" as a "secondary personality" of Pearl Curran. She is also a particularly vivid literary creation. Unfortunately, this makes things more difficult for survivalists, as Braude has rightly pointed out: If a living person can subliminally acquire the information it needs to effectively impersonate another personality, dead or non-existent, then our confidence in even the best survival cases is weakened.

We may still look at the case of Patience Worth as a remarkable piece of evidence for a distinct domain of study: mediumship in service to the creative advance of personality itself. "Patience Worth" was a distinctly original personality of unusual creative capacities. She dictated a 60,000-word poem called *Telka* in 35 hours amid distractions, talking, and writing with others on other subjects. As a linguistic object, this poem has unusual properties. "It is the locutions of *Telka*, and in particular the vocabulary and the uses of that vocabulary, that are, I believe, miraculous" (Prince, 1964:357). For one thing, 90% of the text is in Anglo-Saxon, which is very difficult for most writers and periods of English literature (except possibly Wyckliffe's *Bible*). The language of this mediumistically produced poem is consistent with a composite of 17th century dialects. More to the point, Yost says: "It [the language of *Telka*] seems not to be the language of any period of England nor of any locality of England." It is a language that apparently has never been written or spoken anywhere.

According to Yost, "it is her construction, more even than her vocabulary, that is peculiar and individual. She has no reverence for parts of speech, and little respect for the holiest rules of syntax. 'I be dame,' she says." Take another example from *Telka* where Patience turns the adjective *soggy* into a verb *soggeth,* adding archaism to syntactical transposition. Yost points out that *soggeth* is recorded nowhere else in the history of the English language; it is a pure child of Patience.

Caspar Yost sees this singularity of linguistic performance as proof of the uniqueness of the Patience Worth persona. With regard to the syntactical originality of *soggeth*, it's important to note that originality by itself does not constitute creativity or indicate genius. Syntactical originality, though a possible feature of literary creativity, is neither sufficient nor necessary. Creativity, or genius, in a full-blown Myers-like sense, demands more than oddness or cleverness. In fact, the sentence in *Telka* with *soggeth* is pretty flat writing.

But this is not to say that Pearl Curran was not in her own way highly creative, if not, in a nonliterary sense, a "genius." Her stunts of composition, though impressive, seem to me to fall short of meriting the qualifier creative and seem better described as anomalous. The creative aspect of Pearl's mediumship I want to call attention to is not literary or esthetic but involves another domain of creative intelligence. The creativity here is therapeutic and existential, and consists of Pearl experimentally producing an original and unconventional self; wittier, bolder, earthier, wiser, playful, coyly archaic, and all decked out in Anglo-Saxon idiolect and syntax. Pearl creates Patience, an alternate self who becomes a famous literary performer. It is worth noting that gradually Pearl starts to become Patience, as Prince (1964) reminds us in his account. Becoming this new personality thus exemplifies a domain of creative performance.

The creative act here seems to consist of Pearl subliminally producing a new point of view, a new organizing center for her consciousness. The new center somehow allows new capacities to emerge. When Pearl becomes Patience, she marshals the information she needs, perhaps paranormally, to synthesize her new personality and ratify her story. Prince (1964) concluded that either survival or a hugely expanded view of the human subconscious were the two possible interpretations of Curran's literary output.

Hélene Smith: The Medium as Performance Artist

The career of the medium Hélène Smith (nee Catherine Elise Muller) was especially rich in this special type of creative performance. There are two detailed studies of her life and work (Deonna, 1932, Flournoy, 1899). I will not discuss her reputed paranormal talents. Despite Flournoy's carefully argued criticisms of them, she did (if reluctantly admitted by Flournoy) display some extremely puzzling capacities. Smith's automatic drawings and paintings are quite striking, apart from the totally unconscious way she produced them. Deonna (1932) has reproduced many of these in his massive study of her automatic art.

Smith described how she produced her artworks in a statement to *Light* (October 11, 1913):

> On the days when I am to paint I am always roused very early—generally between five and six in the morning—by three loud knocks at my bed. (Raps or "typtology" was part of her mediumistic repertoire.) I open my eyes and see my bedroom brightly illuminated, and immediately understand that I have to stand up and work. I dress myself in the beautiful iridescent light, and wait a few moments, sitting in my armchair, until the feeling comes that I have to work. It never delays. All at once I stand up and walk to the picture. When about two steps before it I feel a strange sensation, and probably fall asleep at the same moment. I know, later on, that I must have slept because I notice that my fingers are covered with different colors, and I do not remember at all to have used them. . . . (Fodor, 1966:350)

Smith apparently painted without a brush, proving herself a finger-painter extraordinaire. Her paintings, as far as I can see in reproductions without color, are well-composed, smoothly executed with defined images that exude a surreal religiosity that compares favorably with the paintings of Frida Kahlo.

Other aspects of Hélène Smith's creative performance include her myth-making, or mytho-poetic, talents. She was, to begin with, a physically imposing person, tall, beautiful, and highly intelligent. Beyond her everyday functional identity, Catherine Muller became Hélène Smith: medium, clairvoyante, sensitive, automatist, table-tilter, and all-round communicator with Otherworlds. Hélène claimed to communicate with Mars, learned to speak Martian, and invented a written Martian language; she produced visionary paintings of Martian cities and landscapes. Her performance was extremely effective, and the dark-haired beauty with black eyes entranced many followers.

The more circumspect Flournoy, however, succeeded in showing how the literal content of Smith's performances, her claims—being the reincarnation of Marie Antoinette, in touch with Cagliostro, talking with Martians—were false. Hélène and her followers were outraged by Flournoy's deflating critique and punished him by refusing to continue research with him.

Hélène was no doubt taken in by the intensity of her own psychic powers, which did occasionally, but not consistently, bend the parameters of ordinary reality. It's hard to imagine receiving such massive amounts of subliminally mytho–poetic uprush. Her immediate psychological reality must have been subjectively overwhelming, or at least incredibly rich.

The literal failures of her assumed identities took nothing away from Hélène's creative capacity in the special domain we are focusing upon here.

We admire a writer who creates an interesting or entertaining character; as we do an actor who enacts an invented or improvised role. The art of Hélène is uniquely personal, intimate, and integral. As a performance artist, she used her mediumistic talents and mytho–poetic imagination to create *for herself* a series of roles, stories, and adventures; she performed these roles by living, breathing, and identifying herself with them. She used them to expand her own identity, thrusting into the background her old, relatively mundane persona, and making herself over into a new personality, and one that was larger than life. But to succeed she must convince her audience, which she needs to validate her new reality. Hélène was a pioneer in the domain of personality-creation, of what Keats called "soulmaking" and Jung called "individuation."

Matthew Manning: Mediumship as Creative Evolution

Hélène Smith was in some ways like the contemporary Matthew Manning. Both were versatile and did automatic writing, drawing, and painting; developed their capacities as time went on; and were driven by a sense of purpose. Deonna (1932) discusses what he calls Smith's belief in her "mission."

Manning is more critical and consciously experimental in his approach to his ostensibly psychic endowments. He combines active intelligence with creative channeling in the arts and published an account of his psychic evolution. Several points in his narrative bear on the problem of creativity. For one, he seems to have succeeded in redirecting disturbing emotional forces into automatic writing, forces that had at first expressed themselves as teenage poltergeistery. He was in school writing an essay when he suddenly felt something seize his hand. Later he wrote:

> If it looked as though disturbances were imminent, I would sit down and write. Later, it became clear to me that the writing was the controlling factor. It appeared that the energy I used for writing had previously been used for causing poltergeist disturbances. (Manning, 1974:144)

Manning writes that he developed telepathy and a sensitivity to auras and electrical forces. These strange forces about his person occasionally misbehaved and bent keys or did other mischievous things. He also began to tune into spirits, in particular the ghost of Robert Webb from the 18th century, a spirit who still felt like he owned the house that Manning was living in. Sometimes a notable excarnate such as Bertrand Russell dropped in and delivered a message, which Manning did not take at face value. Manning produced hundreds of automatic drawings and paintings, claiming to emanate from great artists such as Durer, Goya, and Picasso. What I have seen of his Goya-like productions seemed authentic.

It seems unlikely that these drawings, done in the styles of the original artists, were indeed the products of their deceased agency, and anyway we are more concerned with *Manning's* creative performance. It is as though Manning, at this stage of his psychic evolution, went from being a poltergeist agent to a kind of open mike for the spirits of great artists to express themselves in the present. The story of Matthew Manning illustrates the growth potential of mediumistic creative powers. Manning, moreover, unlike Hélène Smith, appears more prone to combine creative mediumship with critical thinking. The critical attitude toward his own psychic talents, as it seems from his narrative, served to drive (not inhibit) their creative development. This of course must be a delicate operation; there is the risk that too much critical intervention may impede the flow of intuition.

Creative Cooperation

In 1923–1924, Hester Travers Smith and a Mr. V. jointly produced 25 scripts by means of automatic writing that claimed to derive from the deceased mind of Oscar Wilde. Neither had any particular prior interest in Wilde; Mr. V. was mainly looking to learn automatic writing.

The writing style of the scripts was clearly suggestive of Wilde's clever epigrams and startling turns of phrase. The script itself strongly resembled Wilde's handwriting and contained references to his life that were consciously unknown to the automatists. This interesting material, according to Eleanor Sidgwick's (1924) account, falls short of supporting the survival of Wilde, but more to our purpose supports the idea that mediumistic, or dissociated, states of mind may on occasion enhance creative performance. In this case, moreover, the creativity seemed to depend on a special cooperative relationship.

V. was not altogether successful in learning to write automatically, and it was not until Hester rested her hand lightly on V.'s hand that the scripts came. It is not clear what the guiding intelligence was, or if there was one intelligence responsible for producing the scripts. It does seem that the partnership or convergent dynamic of V. and Hester enabled them to do what they could not singly do. Clearly, some latent capacities require group support before productivity becomes possible.

Mr. V. was not aware of having seen Wilde's handwriting prior to his automatic performance, but the possibility cannot be ruled out that he did, perhaps only half consciously in a bookstore. If we knew he did, it would detract from the survival hypothesis, but not from the creative value of the attempt to produce a convincing persona of Oscar Wilde. The scripts revealed a knowledge of astronomy that Wilde was not known to possess but that Mr. V. did possess. This would seem to show it came not from the deceased Wilde but from V.

It also reveals the opportunism of the creative process. Mr. V. subliminally draws upon his own unconscious memories of astronomy to flesh out the portrait of the character he is creating. This happens in other creative activities; fiction writers draw upon personal experience to create literary characters or imaginary portraits. Mr. V.'s subliminal intelligence apparently slipped up here, as far as consistency in portraying Wilde.

Impersonating the Dead

Finally, consider the most extraordinary, and perplexing, form of mediumistic creativity. Like fine actors who convincingly impersonate fictional characters, superb mediums such as Leonora Piper and Gladys Osborn Leonard have been able to convince their "audience" that they have truly impersonated—or transmitted—personalities known to be deceased.

In order to accomplish this startling effect, they apparently draw upon the full resources of their subliminal selves, psychically scanning the environment for the information needed to recreate a verisimilar deceased personality. Good mediums can reproduce voice, tone, mannerisms, characteristic turns of phrase, and other personal traits of a deceased person so that the sitters of a seance are duly amazed, may be intellectually persuaded, and are sometimes deeply moved.

Both interpretations are evidence for high mediumistic creativity. Suppose the medium has in fact succeeded in being possessed by an identifiable excarnate personality; such a performance may well seem creative. It involves the synthesis of many skills, some of them paranormal. In part, what is created in the audience is a sense of emotional and rational conviction that so and so has indeed survived death. On the other hand, we might equally suppose the medium has only created a convincing facsimile, a compelling illusion of a deceased person. This seems no less a creative feat, raising deception to a paranormal art. This prima facie function of mediumistic performance—producing conviction of the reality of survival—ranks as something we should honor with the title "creative." As such, mediums such as Leonora Piper and Gladys Osborn Leonard deserve to be ranked as geniuses in their domain.

The Artist as Medium

People primarily thought of as mediums display various creative behaviors. Now consider examples of some poets with mediumistic tendencies. The focus on poets is arbitrary but has an advantage: Poets are well-suited to the task of describing the subtleties of their psychical lives.

But first a glance at the Muses, imaginative figures deeply embedded in Greek mythology. In Hesiod's *Theogony* (Evelyn-White, 1959:181), we read:

"And one day they taught Hesiod glorious song while he was shepherding his lambs under Helicon." The muses declared: "Miserable wretches, mere bellies, we know how to speak many false things as though they were true; but we know, when we will, to utter true things." And then they "breathed into me a divine voice. . . . " This last phrase illustrates the idea of *inspiration*—being "breathed into" from an external source. The Muses (goddesses, nine in all, and daughters of Memory) symbolize this external source.

The text shows there is a discontinuity between poet and source of inspiration; in terms of Myers' psychology, a discontinuity between supraliminal and subliminal strata of mental life. Also, Hesiod divined a fact that Myers would later confirm: the subliminal mind is both "evolutive" and "dissolutive," a source of true insight but also of rubbish and prevarication.

The Muses are ingrained in Western literature. From Homer to Milton we find their invocations, an indicator of the ties between mediumism and creativity. In Plato's *Phaedrus,* "the greatest blessings" of healing, music, and poetry arise through "god-given madness." Plato preserves Hesiod's and Myers' distinction between kinds of manic inlet to the subliminal: the god-given "evolutive" and the ordinary "dissolutive." The functional meaning of the Muse was rediscovered during the Renaissance, the Romantic period, and by 20th century Surrealists.

This recognition of genius as something initiated from outside personal mental life was a popular conceit of the Romantic poets. In one of his letters, John Keats remarked on the psychology of creativity:

> . . . at once it struck me, what quality went to form a Man of Achievement especially in Literature & which Shakespeare possessed so enormously—I mean *Negative Capability*, that is when man is capable of being in uncertainties, Mysteries, doubts, without any irritable reaching after fact & reason—. . . (Gittings, 1970:43)

What is "negative capability" that Keats thinks defines creative achievers? It is an ability to escape the constraints of Myers' "supraliminal" mind, the part of us that "itches" to reduce things to their factual and rational character, in short an essentially unpoetic, uncreative way of relating to the world. The poet with negative capability is open to all the possibilities and can tolerate, and even prefer, uncertainty, ambiguity, mystery, and doubt, which well describe the character of the subliminal world, the subwaking region of ideas, dreams, and visions.

Other poets have given away precious secrets in their letter writing. The boy-wonder French poet Arthur Rimbaud left a letter with his theory of "creative dissociation" (Grosso, 1997). Rimbaud describes his method for becoming a seer and clairvoyant; based on a *dereglement*, a "de-ruling" (the opposite of what computers do), a derailing or disordering, of our ordinary sense modalities. Rimbaud strove to retain the esthetic essence of extreme experiences and to dismantle the routines of the conscious ego. He sums up his view with the laconic formula: *Je est un autre* or "I am another" or "there is another me." There is, in short, more to the poet than his supraliminal self. Rimbaud's poetic method was based on systematically disrupting the workings of the supraliminal mind.

More than one great poet has been known for his mediumistic inclinations. Consider the visionary, mediumistic, revolutionary, and extrovertive mystic William Blake, who wrote: "I must create a system or be enslaved by another man's; I will not reason & compare; my business is to create" (Keynes, 1972:629). Creativity for Blake was the necessary expression of his freedom and a response to the dictates of inspiration. Blake wrote his friend Butts that his prophetic poem *Milton* was written "from immediate dictation, twelve or sometimes twenty or thirty lines at a time, without premeditation, and even against my will", and later added, "I may praise it, since I dare not pretend to be other than the secretary; the authors are in eternity" (Damon, 1958:202).

According to Damon (1958:201), Blake "struggled his whole life through to join the Conscious and the Subconscious." This is what Myers understood to be the essence of genius. Blake was conscious of responding to the dictates of the subliminal when he wrote: ". . . I am under the direction of Messengers from Heaven, Daily and Nightly" (Damon, 1958:205).

The life of genius is not without cost. "But if we fear to do the dictates of our Angels," Blake declares, we will suffer "dismal torments" and be shunned as traitors in "Eternity." Eternity for Blake is the afterworld, the world of imagination. Blake's refusal to bury his talents has implications for psychotherapy; we do not care for our souls when we bury our talents, and we pay an emotional price when we fail to create our own "system." Inspiration not acted upon is a sin against the Holy Ghost; as he puts it in one of his Proverbs of Hell: "Sooner murder an infant in its cradle than nurse unacted desires."

Our next example of mediumizing genius is William Butler Yeats, who greatly admired William Blake. Like Blake, Yeats, as he writes in the opening pages of *A Vision* (Yeats, 1962), is driven to use an often strange and barbaric symbolism with the express purpose of uniting "the sleeping and waking mind." Effecting a union of the sleeping and waking mind is exactly what Myers means by genius.

Yeats entertained a theory of magic based on the "the great memory," the transmarginal or subliminal memory (Yeats, 1961). The "magic" consists of using symbols to evoke the powers of the great memory, and thus connect, coordinate, and unite our sleeping and waking mental regions. Yeats's idea of magic and Myers' of genius converge.

In *A Vision,* Yeats declared that his recent "poetry has gained in self-possession and power" and adds, "I owe this change to an incredible experience." The incredible experience was that on October 4, 1917, his wife Georgie Hyde-Lees surprised him by attempting automatic writing. Profound and exciting utterances came forth, and an unknown writer (or writers) said: "We have come to give you metaphors for poetry." Thus commenced an extraordinary partnership in creativity that Yeats pursued with his wife for three years. "Exposition in sleep came to an end in 1920, and I began an exhaustive study of some fifty copy-books of automatic script." These copybooks were the raw materials from which he produced some of the greatest poetry and prose of the 20th century.

Recall the case of V. and Hester in the production of the "Oscar Wilde" scripts: It was a joint effort; when Hester rested her hand on V., the brilliant literary persona of "Oscar Wilde" emerged. Note the analogy with Yeats and his wife, though not exact because Yeats's wife wrote without her husband's hand on hers. But the script was the product of a joint effort, transcending them both, who were more like secretaries to the psychological entity who they jointly produced. The performance was a partnership in dissociation, a species of joint mediumship.

My last example of this is one of the great 20th century American poets, James Merrill, and in particular his masterpiece *The Changing Light at Sandover* (Merrill, 1993), also produced jointly, through a ouija board by himself and his long-time friend. Harold Bloom wrote: "I don't know that *The Book of Ephraim,* at least after some dozen readings, can be over-praised, as nothing since the greatest writers of our century equals it in daemonic force. . . " In the opening pages of that book, Merrill (1993) describes the instant the ouija board comes to life:

YES a new and urgent power YES
Seized the cup. It swerved, clung, hesitated,
Darted off, a devil's darning needle
Gyroscope our fingers rode bareback
(But stopping dead the instant one lost touch) (p. 102)

So, as with V. and Hester, and Yeats and his automatist wife, we note another illustration of creative partnership. If either Merrill or his partner took his finger off the cup, it stopped dead. But when the poet and his companion were one in spirit, results were electrifying.

Yet even the most fragmentary message—
Twice as entertaining, twice as wise
As either of its mediums—enthralled them. (p. 103)

By establishing a certain rapport between individuals, it may be possible to draw on sources of genius otherwise impossible or elusive to employ. Sometimes creative performance seems possible only in the right partnership, the right group dynamic. The subtle group dynamics of the creative process merits further study.

Myers' Theory of Genius

So much for a rapid sketch of two types of empirical linkage between mediumship and creativity. Along the way, I have made several references to Frederic Myers, whose theory of genius (Myers, 1903(1):70–120) provides some useful guiding ideas. Genius occupies a high place in Myers' view of human personality, constituting the "true normality" of the species. The basic intuition is that "genius" is what happens when the "supraliminal" and the "subliminal" strata of our mental life combine, and give birth to new and often useful and illuminating arrangements of ideas and actions. Genius, according to Myers, is the "co-operation of the submerged and the emergent self" and a person of genius "effects a successful co-operation of an unusually large number of elements of his personality" (1903(1)72). Genius implies "co-coordinating the waking and sleeping phases of (one's) existence" (1903(1)73).

Genius is not just about inspiration, however; enlarged receptivity to the submerged self is not the whole picture. Myers puts the stress on volition. The new elements of personality made available to us by inspiration must be shaped and molded by intentionality. "The differentia of genius," he wrote, "lies in an increased control over subliminal mentation" (1903(1):74). Producing works of genius requires patient, steadfast attentiveness—a well-honed will. Genius is a paradoxical state that combines "subliminal uprush" with "increased control" of the subliminal.

Myers' theory enlarges the scope of the activities of genius. The spirit of genius embodies a trend toward greater integration, "a power of utilizing a wider range than other men can utilize of faculties in some degree innate in all" (1903(1):71). Myers here democratizes the capacity for genius; we all possess the germ and principle as part of our innate human endowment.

We associate genius with conventional domains such as art, religion, science, and philosophy, but Myers increases the range of possible experience: ". . . genius may be recognized in every region of thought and emotion. In each direction a man's everyday self may be more or less permeable to subliminal impulses" (1903(1):116).

Every moment of mental life whose superficial character is more or less adapted to the demands of the physical environment also possesses a subliminal depth that is normally hidden from us, although we may count on it being alive at the margins of waking life. What Myers called "subliminal uprush" is possible any moment in any mode of experience: emotional, intellectual, or moral. Raw material is everywhere for genius to work upon.

For Myers genius is an evolutionary process. Reacting strongly to Darwinism, Myers strove to be broadly compatible with the evolutionary outlook, believing that there was a *nisus* or drive in us toward realizing a wider, more comprehensive consciousness. This for Myers was probably rooted in his classically nurtured, neo-Platonic intuitions, and by collating all the relevant empirical data he made it a well-grounded hypothesis. Myers admitted to "enthusiastic" leanings, sensing an upward evolutionary arc at work, in a spirit akin to Aurobindo or Teilhard de Chardin, who emphasize the development of our mental and moral faculties.

Unfortunately, there is little evidence that the species as a whole is utilizing its genius potential; when progress occurs, it does so in isolated, unpredictable individual cases, and occasionally concentrates in times and places.

Myers' theory of genius puts us in touch with phenomena of heroic sanctity, mysticism, and supernormal faculty. Mediumship, as a form of dissociation from the habits of ordinary consciousness, may increase receptivity to these higher and more expansive domains of experience. According to Myers, the fully realized humans of the future will enjoy mastery of the multiple strata of our submerged, latent capacities, and in fact during the 20th century developments continued the legacy of a Myers-like evolutionary psychology.

Surrealism

My first example of a new domain of creative performance is Surrealism. Surrealism, and its chief theoretician Andre Breton, is generally thought to have been inspired by Freud; Myers, however, was also a significant source. In 1933 Breton published *The Automatic Message* expressing Surrealism's indebtedness to the "gothic psychiatry" of F. W. H. Myers (Breton, 1997). Myers' work on automatism provided the psychological mechanism that Surrealists would exploit in novel ways. "Surrealism has above all worked to bring inspiration back into favor, and we have for that purpose promoted the use of automatic forms of expression . . . " (Breton, 1997:16).

Why promote these forms of expression? Breton had little interest in their otherworldly applications, what he calls the soul "dissociating" from the body. He was more interested in transforming premortem experience: ". . . the Surrealists' aim is nothing less than to unify the personality" (Breton, 1997:17). This of course is precisely what Myers meant by "genius," the interpenetration and ultimate coordination of dream and waking life. This is how Breton puts it in the 1924 *Manifesto of Surrealism*: "I believe in the future resolution of these two states, dream and reality, which are seemingly so contradictory, into a kind of absolute reality, a *surreality,* if one may so speak" (Breton, 1972:14).

Surrealism was based on the assumption that it is possible to forge new psychophysical realities by fusing the waking and subliminal strata of our mental life. Surrealism was a school and theory of art; it was also a political program committed to creating a new social consciousness that would be deeply in touch with its dream life. Surrealism was one way that the project of creative mediumship branched out into the 20th century.

Outsider Art

Another, related domain of dissociative creativity is "outsider" art (Hall & Metcalf, 1994). This umbrella term covers the work of children, primitives, the incarcerated, the elderly, folk art, *art brut,* psychotic art, and generally all forms of art and image-making produced by the untaught, the culturally deprived, the isolated, and the marginalized. We may characterize these forms of spontaneous expression as subliminal; it is the creativity that occurs in the face of denial of supraliminal intelligent resources. Outsider artists are less inhibited and more like automatists than conscious craftspeople.

Sometimes creative automatism is like a seizure. Scottie Wilson ran a used goods store in Toronto. "One day in the back of his shop, he picked up a fountain pen and began doodling on a tabletop. He drew until the entire surface was covered, and he continued to draw for the next thirty-seven years of his life" (Petullo, 1993:105). Wilson's work resembles but is more dazzling than Paul Klee's; eventually he had shows in London galleries but refused to sell his artworks. He did succeed, however, in eking out a living charging people to look at his artworks. Scottie kept drawing with colored pencils until his late 70s when he died. It certainly appears as if he were seized by something starkly dissimilar and external to his 35-year old personality, and that it transformed him, driving him to a surprising artistic career.

The case of Augustin Lesage (1876–1954) also illustrates the phenomenon of possession by the "muses." Lesage was a coal miner living in Northern France, a man without anything in his experience or background that could account for his fate as an artist. Sometime around 1912 Lesage had an extraordinary experience. He writes:

> I was in the mine, in a long tunnel, I was working by myself . . . —suddenly, I hear a voice talking to me. I look around on all sides of me . . . I was alone. I was stupefied! I was afraid, and my hair stood up. Then I heard: "Don't be afraid, we are near you. One day you will be a painter". (Notter et al., 1988:12)

The voices continued to urge him to paint and persuaded him to order art supplies from a shop in nearby Lille. To his consternation he received one piece of canvas 10 feet square and a few very small brushes. The voices forbade him to cut the canvas up. Under the guidance of his voices, he commenced his first canvas. He began without drawing or plan in the upper right hand corner and a year later completed his first artwork, in which he combined miniaturist technique with massive symmetrical architecture, which requires high degrees of careful and directed control. Lesage, like William Blake, claimed no responsibility for these works, which have grown in fame. "A picture comes into existence detail by detail," he said, "and nothing about it enters my mind beforehand. My guides have told me: 'Do not try to find out what you are doing.' I surrender to their prompting. . . . I follow my guides like a child" (Maizels, 1996:56). As Lesage became more famous and more sophisticated, he seems to have declined in his ability to deploy his subliminal muses.

Adolf Woelfli (1864–1930) is well-known in the field of outsider art (Spoerri, 1997). In some ways, Woelfli shares with Hélène Smith a profile of extraordinary multi-modal creativity shaped by persistent automatisms. With a Blakean compulsion not to be "enslaved" by others, Smith and Woelfli were driven to create vast, autonomous worlds. Their outer circumstances differed markedly: Woelfli was an institutionalized paranoid schizophrenic for most of his life; Hélène Smith was a vital, attractive woman who made an impressive success of her psychic talents.

Woelfli, given to violent acts and sexual aggression, spent much of his time in solitary confinement at the Waldeau Psychiatric Clinic in Berne, Switzerland, and was pronounced "mentally incompetent." He was institutionalized in 1895 for sexually molesting a young girl because (as he explained) the parents of his true love rejected him on account of his inferior social class.

Woelfli's childhood was beset by poverty and parental ignorance; he lost both his parents who had turned to a life of crime. He was farmed out to State institutions during a low point in the history of Berne. His behavior was violent, aggressive, and he heard voices (paranoid delusions) during the early stage of incarceration. Woelfli was severely cut off from the external world. In 1899, he spontaneously began to write and draw. Encouraged by staff to continue doing so in the clinic, these activities seemed to channel his violent impulses. (Recall that Matthew Manning channeled his destructive poltergeist into automatic writing and drawing.)

Woelfli's productivity was enormous. From 1908 to 1930 he worked on a massive narrative that is a mythology of Adolf Woelfli, a mixture of authentic personal history and cosmic fantasy, a carefully unified whole, woven together with prose, poetry, illustrations, and musical compositions. This mentally incompetent madman left behind him 45 volumes, 16 notebooks, altogether 25,000 packed pages, along with hundreds of drawings that now hang next to the work of Paul Klee in Switzerland.

Notable is the enormous measure of precise control over all this material, reflecting Woelfli's will to create a secondary world for himself in the face of the devastation of his given world. The theme of this epic outpouring of obsessive genius is the constant destruction and resurrection of the central protagonist who is the author himself. It is about survival in this world, about molding with Renaissance gusto the form and image of one's own existence. He accomplished all this with a few pencils and paper doled out by the State while forced to inhabit day in and day out the same small cell.

Morgenthaler (1921/1992) was fascinated by Woelfli's relentless creative output and observed the inmate at his worktable while he drew and wrote. Woelfli would talk about the content of work, but he had no idea how he produced it. Morgenthaler observed that Woelfli worked from no previous sketch and that he drew spontaneously. Despite this, the entire ongoing oeuvre seemed guided by a single compulsive unifying intelligence (MacGregor, 1989:215). Woelfli's performance illustrates what Myers meant by creative subliminal intelligence. Myers, unlike Lombroso and Freud, would judge Woelfli's performance as evidence for an innate capacity for creative process.

Morgenthaler and Hans Prinzhorn rejected Lombroso's insistence on pathologizing genius (Lombroso, however, was a pioneer collector of the art of the insane) and were the first to emphasize the esthetic, therapeutic, and psychological value of the creativity of mentally disturbed patients. In this they confirmed Myers' inclinations, for Prinzhorn, like Myers, saw the creative imagination, what he called *Bildnerei* and *Gestaltung*, "imagery" and "figuration," as an irreducible creative human capacity.

The German investigators felt they had identified a basic property of consciousness; an inborn capacity of mind that is independent of linear, computational intelligence: a strictly internal process "not subordinated to any outside purpose," according to Prinzhorn, "but directed solely and self-sufficiently toward their own self-realization." All expressive gestures, all figuration, in any medium, aim "to actualize the psyche and thereby to build a bridge from the self to others." The purpose of figuration, or the creative imagination, is to enable us "to escape into the expanse of common life from the restrictions of the individual . . . " (Prinzhorn, 1995:13). Creativity is thus essentially tied to the making of human society, and sometimes in psychotic states creative urge works toward healing.

Creative Mediumship and "Soulmaking"

So far we have focused on certain conventional forms of creativity such as poetry and the plastic arts. But Myers' theory of genius allows for subliminal uprushes to occur in a wide range of experiences; in effect, any state, situation, or mode of experience is more or less "permeable" to inspiration. Myers calls us to survey and prepare to explore the greater arena of creative process.

To indicate this briefly, consider the examples of Socrates and Joan of Arc, two creative giants, one in the domain of wisdom, the other of extraordinary leadership. Socrates, the archetypal philosopher and martyr to Western rationalism, throughout his life is guided, in big and small matters, by a daimon, or inner voice, a creature of subliminal intelligence (Myers, 1888–1889:522–547). We have no direct, written statements from Socrates about his inner life, but in the *Apologia* Plato represents Socrates as saying that the daimon was silent about his being condemned to death, from which the philosopher inferred that death could not be a bad thing. In short, even in dying Socrates was consciously attuned to his subliminal self and was guided by it. What makes Socrates so extraordinary is that he seems to have perfectly fused his conscious critical intellect with his subliminal daimon. In the vast majority of human beings, the two are almost always thoroughly disjointed and disconnected, often at great emotional and spiritual cost.

Or take another highly exceptional being, Joan of Arc, the virgin teenager who turned the tide of the Hundred Years War and set France on the road to nationhood. Since Joan was a child in Domremy, and throughout her brief life, she was guided by subliminal intelligence. In Lang's (1895) account of Joan's voices, again we find a case in which the superficial awareness of everyday life is interfused with subliminal messages.

All the major events of Joan's life were influenced by the voices (sometimes accompanied by lights and visions of the saints). She first heard voices when she was 13, at first telling her to pray and go to church. She referred to them as "my voices," "my counsel." Soon they began to prod her about her mission to save France. They helped her gain the confidence of the Dauphin by assuaging secret fears he had about his legitimacy (Lang, 1895:199). D'Aulon, her companion throughout her career, testified that Joan told him she was being "counseled" by her voices on military tactics and general strategy.

In the course of difficulties, her voices encouraged her to "bear all cheerfully" and to believe that her "martyrdom" would lead to "paradise." Obliquely, they predicted the exact time of her death. The voices came to her spontaneously, and if they failed her she summoned them with prayer. She claimed they woke her from her sleep and kept her company even in the din of court proceedings as she defended herself against the wiles of her accusers. When she temporarily recanted her mission, her voices encouraged her to recant her recantation, and do her duty. As in the case of Socrates, the subliminal voices seemed to place integrity of purpose above life itself. It is clear from the records that Joan's surface life was in deep dialogue with her subliminal will—her "voices," her "counsel." The genius of Joan of Arc served a nation's emancipation, and seemed to embody collective, not personal, needs.

We need to recall the fact that Joan grew up hearing widespread rumors about a young maid coming to rescue France from the English oppressors. In fact, there were predecessors, models of mystical and prophetic women known to play roles of leadership in the affairs of state, including, for example, Hildegarde of Bingen (1198–1279), Elizabeth of Schonau (1128–1164), Mechthild of Magdeburg (1210–1297), and Catherine of Siena (1347–1380). All these figures may have played a role in mobilizing Joan's subliminal self. This is conjecture, but it is based on enough empirical data to make it at least plausible. It is conceivable that a purely random suggestion, or series of suggestions from Joan's environment, may have helped crystallize a disposition in Joan to save her country from its English occupiers. In an unrepeatable and unintended experiment, all the right elements may have come into place, which drove and guided her extraordinary exploits.

But let us return to less grand and more down-to-earth examples of creative performance. The restless, ever opportunistic, nisus of genius also expresses itself in personal venues. There is one area of performance we do not usually associate with conventional notions of creativity. And yet this could prove the most interesting form of creative mediumship: namely, in the formation of human personality, the project, as Rimbaud said, of coming "into the fullness of one's dream" (Rimbaud, 1957:28). To put this in perspective, the historian Burckhardt suggested the Renaissance as a time when people began to think of their *lives* as works of art, as a domain of self-perfection; the sentiment is typified by Aretino's motto to "live resolutely" and aim to become, as Castiglione put it, a "universal human" (Burckhardt, 1935:147–150).

More systematic notions of the ethos of self-creation and creative authenticity emerged among 20th century existentialists, particularly Heidegger and Sartre. For our purpose, a famous letter of John Keats to his brothers in 1819 on "soulmaking" will serve to clarify the main idea. Keats had been reading and musing on the fact that the world is a place beset by troubles, a "vale of tears." No matter what obstacle we think we have overcome, he says, we are still likely to encounter "a fresh set of annoyances." He writes: "Call the world if you Please 'The Vale of Soulmaking'" and adds: "There may be intelligences or sparks of divinity in millions—but they are not Souls till they acquire identities, till each one is personally itself." This can only come about through "the medium of the world." "Do you not see how necessary a World of Pains and troubles is to school an Intelligence and make it a soul?" (Gittings, 1970:250). Keats, like Myers, Breton, Yeats, Prinzhorn, and Jung, saw what he called "self-identity" as the goal of "soulmaking"—the integrated embrace of experience; the fusion of opposites into something original and self-realized.

There are many ways to mine the Keatsian idea of "soulmaking." Braude (2002) has shown in what sense sustaining an induced negative hallucination, which is a kind of selective dissociation, puts creative demands on the subliminal intelligence. To tell one lie to yourself effectively, you must tell many more interrelated lies, and that calls for creative ingenuity, at least if you hope to maintain your fiction long-term.

The form of creativity that Braude has analyzed may be extended to apply to a wide range of behaviors. We say, for example, that a person has a "blindspot" about one thing or another; fanatic physicalists, to give an appropriate example, simply cannot see the mountains of psi-supportive data before them. It seems to me that negative hallucinations of a cognitive type are rampant in psychological life. There is probably a measure of negative hallucination in every fanatically or perhaps even *too* firmly held position. I would go further and say there must be at any given moment in our lives all sorts of unconscious restrictions on what we permit ourselves to see.

On the other hand, selective omission (more or less subliminal) may be a factor in shaping a healthy, expansive personality as well. Making oneself, shaping one's personality, what the ancient Greeks called *paideia*, or "education," is perhaps the most practical, the most demanding, of the personal performance arts. Consider how we might all profit from inducing in ourselves various "negative hallucinations": for example, not to think of discouraging thoughts when they are counterproductive, or allow oneself to be hampered by corrosive doubts, or frightened by overblown fears, or undermined by envy or jealousy, or rendered ineffectual from endless regret, or paralyzed by too many scruples and inhibitions. In short, if we could learn to mobilize the subliminal scanner to selectively shield us from awareness of the ideas and associations that we decide in advance might derail or needlessly complicate our path, it would be a tool in the art of sculpting our personalities.

The idea of conscious re-creation of one's personality became more than a Renaissance or Romantic fantasy when the early mesmerists began to discover what was called divided consciousness, secondary personality, or artificial somnambulism. One of the first things that became evident was that the secondary personality—that often emerged in response to some crisis—often seemed to possess traits superior to the primary. In a famous example, a peasant with an inflamed lung, Victor Race, a subject of the Marquis de Puysegur's mesmeric manipulations, manifested a personality that was more alert, articulate, confident, and perceptive than the non-mesmerized Victor. (Puysegur is generally credited with having been the first to elicit a "secondary personality" from a mesmerized subject.)

> Though ordinarily a simple and tongue-tied peasant, he would, in the somnambulic state, converse in a fluent and elevated manner, expressing such sentiments of friendship. . . . Victor assumed management of his own case, diagnosing and prescribing for his illness, and predicting its course. More than this: On being brought into contact with other patients, he seemed able to do the same for them. (Gauld, 1995:41)

Victor eventually began to take on the characteristic form of his secondary personality, thus providing a model for how a personality can be re-created from subliminal resources.

A psychological situation was artificially induced that offered opportunities for the re-integration, healing, and even enlargement of Victor's primary peasant personality. Early investigators such as Janet and Binet came to understand the therapeutic potential of their subjects' multiple selves, as others more recently have. For example, Adam Crabtree offers this stunning reflection: "It is my opinion that it could conceivably be therapeutically beneficial to assist the creation of a full-blown personality that embodies mixed elements already existing within the individual in a disorganized way" (Crabtree, 1985:225). I am struck by how well this statement of Crabtree's describes the everyday business of the artist, which is to confront and create order out of chaos.

In Myers' view of genius and the *Gestaltung* of Prinzhorn, we find the idea that there is a creative, transformative principle, or we might invoke the *esemplastic* faculty, a word invented by Coleridge for the imaginative power to form things into unities. Like an artist or scientist incubating, "sitting on" a question or problem, with all the pieces coming together in a flash of inspiration—in a similar way it might be possible to incubate a new personality, and produce what is needed to flourish under a new identity.

Concluding Observations

My purpose has been to widen the scope of the concept of mediumistic creativity. I have stressed what I call the domain of personality-creation. The idea that we possess the power to transform our personalities lies at the foundation of Myers' hypothesis of the subliminal self:

> I hold that we each of us contain the potentialities of many different arrangements of the elements of our personality, each arrangement being distinguishable from the rest by differences in the chain of memories that pertains to it. The arrangement with which we habitually identify ourselves—what we call the normal or primary self—consists, in my view, of the elements selected for us in the struggle for existence with special reference to the maintenance of ordinary physical needs, and is not necessarily superior in any other respect to the latent personalities which lie alongside it—the fresh combinations of our personal elements which may be evoked, by accident or design, in a variety to which we can at present assign no limit. (Myers, 1888:374–397)

According to this remarkable statement, our "normal" self is largely a cut-out from a much greater subliminal self, but a cut-out ideally adapted to biological survival; whereas the potential for other arrangements of our personality factors has no assignable limit. Each of us harbors a multitude of possible centers of new, richer, and profounder selfhood, ranging across all human forms of experience. We have reviewed a small number of examples from the literature of genius and mediumship. The possibilities of human transformation are far richer than mainstream psychology seems prepared to admit.

References

Braude, S. E. (2002). The creativity of dissociation. *Journal of Trauma and Dissociation, 3*, 5–26.
Braude, S. E. (2003). *Immortal Remains: The Evidence for Life after Death*. Lanham, MD: Rowman & Littlefield.
Breton, A. (1972). *Manifestoes of Surrealism*. Ann Arbor: University of Michigan Press.
Breton, A. (1997). *The Automatic Message*. London: Atlas Press.
Burckhardt, J. (1935). *The Civilization of the Renaissance in Italy*. New York: Albert & Charles Boni.
Crabtree, A. (1985). *Multiple Man: Explorations in Possession and Multiple Personality*. New York: Praeger.
Damon, S. F. (1958). *William Blake: His Philosophy and Symbols*. Gloucester, MA: Peter Smith.
Deonna, W. (1932). *De la Planete Mars en Terre Sainte: Art et Subconscient*. Paris: Seuil.
Evelyn-White, H. (1959). *Hesiod: The Homeric Hymns and Homerica*. Cambridge, MA: Harvard University Press.
Flournoy, T. (1899/1963). *From India to the Planet Mars*. New York: University Books.
Fodor, N. (1966). *Encyclopedia of Psychic Science*. University Books.
Gauld, A. (1995). *A History of Hypnotism*. New York: Cambridge University Press.
Ghiselin, B. (1952). *The Creative Process*. New York: New American Library.
Gittings, R. (1970). *Letters of John Keats*. London: Oxford Paperbacks.
Grosso, M. (1997). Inspiration, mediumship, surrealism: The concept of creative dissociation. In: Krippner, S., & Powers, S. M. (Eds.), *Broken Images, Broken Selves; Dissociative Narratives in Clinical Experience* (pp. 181–198). Washington, D.C.: Brunner-Mazel.
Hall, M. D., & Metcalf, E. W. (1994). *The Artist Outsider*. Washington, D.C.: Smithsonian Institute Press.
Keynes, G. (1972). *Blake: Complete Writings*. London: Oxford University Press.
Lang, A. (1895). The voices of Jeanne d'Arc. *Proceedings of The Society for Psychical Research, 11*, 198–212.
MacGregor, J. M. (1989). *The Discovery of the Art of the Insane*. Princeton University Press.
Maizels, J. (1996). *Raw Creation: Outsider Art and Beyond*. New York: Phaidon.
Manning, M. (1974). *The Link*. New York: Holt, Rinehart, & Winston.
Merrill, J. (1993). *The Changing Light at Sandover*. New York: Alfred J. Knopf.
Morgenthaler, W. (1921/1992). *Madness and Art: The Life and Works of Adolf Woelfli* (Aaron Esman, trans.). Lincoln, NE: University of Nebraska Press.
Myers, F. W. H. (1886–1887). Human personality in the light of hypnotic suggestion. *Proceedings of The Society for Psychical Research, 4*, 1–24.
Myers, F. W. H. (1888). French experiments in the strata of personality. *Proceedings of The Society for Psychical Research, 5*, 374–397.
Myers, F. W. H. (1888–1889). The daemon of Socrates. *Proceedings of The Society for Psychical Research, 5*, 522–547.
Myers, F. W. H. (1896). Glossary of terms used in psychical research. *Proceedings of The Society for Psychical Research, 12*, 166–174.
Myers, F. W. H. (1903). *Human Personality and Its Survival of Bodily Death*. 2 Vols. London: Longmans, Green.
Notter, A., Deroux, D., Thovoz, M., & Amadou, R. (1988). *Augustin Lesage 1876–1954*. Paris: Philip Sers.
Petullo, A. (1993). *Driven to Create*. Milwaukee, WI: Milwaukee Art Museum.
Prince, W. F. (1964). *The Case of Patience Worth*. New York: University Books.
Prinzhorn, H. (1995). *Artistry of the Mentally Ill*. New York: Springer-Verlag.
Rimbaud, A. (1957). *Illuminations*. New Directions.

Schiller, F. C. S. (1928). The case of Patience Worth. *Proceedings of The Society for Psychical Research, 36,* 573–576.
Sidgwick, E. (1924). The "Oscar Wilde" scripts. *Proceedings of The Society for Psychical Research, 34,* 186–196.
Spoerri, E. (1997). *Adolf Woelfli.* Ithaca, NY: Cornell University Press.
Stevenson, R. L. (1966). *The Strange Case of Dr. Jekyll and Mr. Hyde.* New York: Dell.
Yeats, W. B. (1961). *Essays and Introductions.* New York: Macmillan.
Yeats, W. B. (1962). *A Vision.* New York: Macmillan.

ESSAY

Some Directions for Mediumship Research

Emily Williams Kelly

*Division of Perceptual Studies, University of Virginia Health System
Charlottesville, Virginia 22902, USA
ewc2r@virginia.edu*

An earlier version of this paper was presented at a Parapsychology Foundation International Conference, The Study of Mediumship: Interdisciplinary Perspectives, *in Charlottesville, Virginia, January 29–30, 2005.*

Abstract—The study of mediums was part of a larger program of psychical research, begun in the late 19th century, intended to examine specifically whether human personality survives bodily death, and more generally whether the brain produces mind or consciousness, as most scientists since the late 19th century have assumed. Although a vast amount of high-quality research resulted from that effort, the study of mediumship was almost completely abandoned during the latter half of the 20th century, primarily because of the impasse reached over whether the phenomena are best-interpreted as attributable to deceased agents or to living agents. In this paper the author examines some types of mediumship research that have been considered particularly important for the survival question: cross-correspondences, drop-in communicators, and proxy cases. She argues that a revival of research on mediumship, particularly with proxy sittings, could contribute importantly to present-day psychical research and, perhaps ultimately, move us beyond the current impasse.

Keywords: mediumship—survival—proxy research—cross-correspondences—drop-in communicators

Introduction

The program of research that came to be known as psychical research, launched in the late 19th century, was motivated by one primary question—namely, whether the assumption that mind or consciousness is solely the product of the brain is adequate to account for all of human experience (Cook [Kelly], 1992, Kelly et al., 2007). For many of the founders and early followers of psychical research, behind this general issue lay the specific and emotionally significant question of whether human personality survives bodily death. An integral part of the research, therefore, was the study of the phenomena that gave rise to Spiritualism, in which some persons, called mediums, seemed to move or otherwise influence physical objects, or to communicate, through automatic writing, planchette, or trance speaking, information of which the medium was not consciously aware. The Spiritualists attributed the phenomena to the action of deceased spirits, but with the exposure of many fraudulent mediums (see, e.g., Gauld, 1968, Chapter 9), together with the growing recognition that much of the information communicated probably came from the medium's own subconscious mind and not from deceased persons (e.g., Myers, 1884), psychical researchers increasingly concentrated their research on a selected few mediums who were able to provide specific and verifiable information about identifiable deceased persons that the medium had no normal way of knowing.

The first 50 years of research on trance, or mental, mediumship produced a vast amount of evidence that led nearly everyone knowledgeable about it to conclude that the mediums obtained the information through *some* supernormal process, although there was considerable disagreement about what that process was (for excellent reviews, see Braude, 2003, Gauld, 1982). In this paper I will briefly review three particular types of evidence from mediumship that many people have considered particularly strong support for the hypothesis that human personality survives death. My primary goal is to suggest some directions that I believe researchers could take now to add to and extend the already impressive body of research on mediumship.

First, however, I would like to make some comments about why I believe research on mediumship is so important, beyond its specific application to the question of survival. It is the only phenomenon in psychical research that combines elements of spontaneous case studies, field studies, the study of special subjects or individuals, survival research, and experimental method. Mediumship usually develops spontaneously in a few gifted individuals; it must be studied more or less in its natural setting; and yet it is the only phenomenon directly relevant to the survival problem that can be produced and observed under conditions of experimental control. Mediumship therefore combines the significant emotional and psychological circumstances that often produce the strong psi effects seen in spontaneous experiences with the ability to control the conditions and thus reduce the uncertainties that too often accompany spontaneous cases with regard to the possibilities for normal explanations or sources of information. In a field in which spontaneous case studies, survival research, and experimental research have become for the most part widely segregated, a phenomenon that brings all these aspects together would seem to be singularly important for pursuing psychical research from a broader perspective.

Moreover, the study of mediumship may provide an important means of addressing variations of a question that underlies all of psychical research and parapsychology[1]: In any given psi experience or event, who is the *source*?

Should veridical mediumistic readings be attributed primarily to living persons (the medium? the sitters?) or to deceased persons? Should veridical crisis apparitions be attributed primarily to the percipient or to the dying person? Should successful results in a psi experiment be attributed primarily to the ostensible subject or to the experimenter? Or, in all of these, is it some interaction among all parties involved?

Keeping in mind the broad relevance and importance of mediumship for psychical research in general, I will now turn to mediumship more specifically as a means of studying the survival question. When evaluating mediumistic material, there are two major components involved. First, we have to determine whether the statements of the medium might have come from some normal process, or whether the most likely explanation is that they have derived from some supernormal process. We are all well aware of this, but it cannot be repeated too often that there are several primary normal processes that must be ruled out before we can say that a supernormal process was involved. First, of course, we have to rule out fraud. I think most mediums are honest, straightforward people whose primary goal is to help people, but we do have to be alert for any indications of fraud. I had a sitting many years ago with a medium who asked that the sitter write down on a card questions that he or she would like to ask deceased loved ones. This is a suspicious procedure in and of itself, and regrettably in this case suspicion was well-founded. Without describing the unfortunate details, I can say that the medium's machinations to obtain the card from me and read it were almost comical; but they are not amusing when one considers how many bereaved people he was deceiving. I think he was the exception rather than the rule among mediums, but there are other normal explanations that may be operating in many instances, and two in particular are the most important to rule out. One of these is fishing for information (or "cold reading"), whether done deliberately or inadvertently. Even when a medium does not allow the sitter to say anything but "yes" or "no" to a statement, a great deal of information and direction can be obtained in this way. The other normal explanation is that the medium makes vague or general statements that could apply to many people. It is only when we have ruled out normal explanations such as these that we can consider a supernormal hypothesis.

If we *do* decide that a medium's statements are most likely the product of some supernormal process, we can then go on to the second step, which is to consider what that process is and whether it involves the participation of a surviving deceased person or whether the information could have come instead from the medium's ability to obtain the information in a supernormal manner from terrestrial sources alone. These competing hypotheses constitute the survival hypothesis and the super-psi (or super-ESP) hypothesis, respectively. From the earliest days of research with mediums in the late 19th century, researchers had recognized that telepathy between living persons was a serious alternative explanation to the deceased-agent hypothesis, and there had been herculean efforts by earlier researchers to identify and study mediumistic communications difficult to account for by a super-psi hypothesis. Three types of mediumistic communications that, for many people, began—even if only slightly—to tip the scale away from a super-psi interpretation and toward a survival one were the *cross-correspondences*, *drop-in communicators*, and *proxy sittings*. The cross-correspondences and drop in-cases were thought to provide good evidence for survival because the motivation for the communications seemed stronger on the part of the deceased person than on the part of any living persons. Proxy sittings were thought to provide good evidence because, without any persons present who knew the deceased person, it seemed less likely that a telepathic "link" had been established between the medium and living people who knew the information contained in the communications.

Nevertheless, as knowledge about psi, including telepathy, clairvoyance, precognition, and psychokinesis, grew, it began to seem that in psi phenomena, the *goal* of the task at hand may be more important than the apparent *complexity* of the task. The medium's goal to provide evidence of survival might be sufficient to elicit the needed information, by clairvoyance or telepathy with living persons, in however complicated a way. Without knowledge of any limits of "normal" psi, it was impossible to say that a medium's statements had crossed those limits and established the survival of a deceased person. It became, and remains today, largely a matter of personal judgment whether one decides that a medium's statements are evidence for psi among living persons or for psi between a living medium and a deceased person. As a result, despite more than 50 years of high-quality research on mediumship, the research more or less ground to a halt, primarily because researchers had reached an impasse when it came to evaluating these alternative explanations (Cook [Kelly], 1987, Gauld, 1961). There seemed to be no way to discriminate between them, or to falsify one and establish the other. Although the general public today is showing a revival of interest in mediumship, primarily because of television shows about, and books by, a few high-profile mediums, few scientists have taken it up again as a topic for serious research.

I would argue, however, that intimidation in the face of the survival/super-psi impasse led to the suppression of one of our most valuable sources of information and phenomena relevant *both* to psi and to the survival question. However one interprets the phenomena, there is little doubt among people who have studied the evidence provided by the best mediums of the 19th and early 20th centuries that at least some of the information given was not obtained in any normal way. Ceasing to identify good mediums and methods for producing the phenomena, simply because the theoretical impasse seemed so difficult to surmount, was, in my view, a serious self-inflicted wound on psychical research. It will only be by the accumulation of more evidence that we may eventually begin to see our way beyond the impasse; and in the meantime, while accumulating this evidence, we may also learn much about conditions conducive to producing supernormal phenomena in general. As one researcher put it: "the accumulation of evidence is . . . a matter of great importance; it is impossible to have too much" (Saltmarsh, 1929:53).

The primary purpose of this paper, therefore, is to encourage renewed research on mediumship which produces evidence, like that from previous research, that *some* supernormal process is occurring. I have no wish to debate the strengths or weaknesses of either the survival or the super-psi hypothesis for any particular case or type of case, for the simple reason that I find myself at the same "impasse" that so many researchers before me have. Moreover, I agree with those who find the supposed dichotomy between living-agent and deceased-agent hypotheses to be less than clear-cut, since psi must be operating in either case. As Michael Sudduth (2009) has so well demonstrated, arguments that undercut living-agent psi are a double-edged sword and work equally well to undercut deceased-agent psi. Like Gardner Murphy (Murphy, 1945a), I doubt that the "questions [about survival] have been rightly stated," and "think it probable that five hundred years hence the arguments both pro and con will sound childish and superficial, if indeed they sound relevant to the problem at all" (Murphy, 1945a:93).

Rather than enter into the theoretical fray, therefore, in this paper I wish simply to discuss the three lines of research that I think are most important for future research with mediums, particularly if we wish to advance our thinking about the survival issue. The cross-correspondences and drop-in cases are important because in them the motivation or purpose seems stronger on the part of the deceased person than on that of the medium or other living persons. Despite their importance, I will discuss these cases only briefly because they occur for the most part only spontaneously. Research with proxy sittings, on the other hand, can be undertaken deliberately, and for this reason it is a more productive method. I also believe it is the most important line of research to pursue now, not so much because it lessens the super-psi hypothesis, but because it lessens the likelihood of normal explanations such as cold reading and biased interpretation of vague statements.

Cross-Correspondences

Cross-correspondences began to develop as a new mediumistic phenomenon shortly after the death of Frederic Myers on January 17, 1901. They purportedly represented an experiment developed on the other side primarily by the deceased Myers but also by his deceased friends and colleagues in psychical research, Henry Sidgwick and Edmund Gurney. All three of them were well aware during their lifetimes of what we might now call the Catch-22 aspect of mediumistic communications: Statements by mediums are only useful for scientific purposes if they are veridical, that is if the information can be verified as true and accurate; but if they are to be so verified, this must be by the memory of a still-living person or by documents or other physical evidence. That being the case, the statements could always be said to have derived somehow from telepathy or clairvoyance. The cross-correspondences were therefore said to be an attempt to circumvent this argument by having two or more mediums—who were of course not in contact with each other—give statements that, when taken alone, made no sense, but when put together did. The argument was that the underlying meaning tying the statements together originated in the mind of one person only, and that was the deceased purported communicator.

There were five primary mediums, and a few others to a lesser extent, involved in the cross-correspondences. One was Mrs. Leonora Piper, a professional medium who had been working with members of the Society for Psychical Research (SPR) for nearly 20 years. The other four primary mediums were private individuals, however, and not professional mediums. Because all of the communications were obtained by automatic writing, they were usually referred to by SPR investigators as "automatists" rather than as "mediums," and I will do so here as well.

None of the five primary automatists were in contact with each other. The cross-correspondences came to light particularly after Mrs. Holland (a pseudonym for Rudyard Kipling's sister), who was then living in India, wrote automatically on November 7, 1903, a message purporting to come from Frederic Myers, who had died nearly three years earlier. The message was addressed to Mrs. Verrall, a friend of Myers's whom Mrs. Holland had heard of but never met, and it ended with the instruction to send the writing to Mrs. Verrall at 5 Selwyn Gardens, Cambridge. Mrs. Holland had never been to Cambridge and had never heard of Selwyn Gardens, but this was Mrs. Verrall's address. From this point on, Alice Johnson and other investigators from the SPR became involved, and cross-correspondences continued for several decades.

There is a voluminous amount of material relating to the cross-correspondences (for an introduction to the cross-correspondences cases and references to the main papers, see Gauld, 1982, Murphy, 1961, Saltmarsh, 1938), and most of it is extremely complex because it is often centered around allusions to classical literature, in which all the three primary deceased communicators, some of the automatists, and many of the SPR investigators were well-versed. One of the main objections to the cross-correspondence material, therefore, is the highly recondite nature of much of it; few people now can study the original automatic writing and put the various pieces together for themselves. But I will describe one of the simpler cross-correspondences, just to give some sense of what was involved.

This material was reported by Alice Johnson in a 1908 paper (Johnson, 1908) on the automatic writing of Mrs. Holland and was called by her the "Roden Noel" case. Roden Noel was a minor poet in the 19th century. On March 7, 1906, Mrs. Verrall (in England) wrote automatically a poem that began "Tintagel and the sea that moaned in pain." This meant nothing to Mrs. Verrall, but she as usual passed on her automatic writing to Alice Johnson.

On March 11, 1906, Mrs. Holland (in India) wrote a message, seemingly coming from the deceased Henry Sidgwick, in which Sidgwick said to ask AW (meaning Mrs. Verrall's husband, Dr. A. W. Verrall) "what the date May 26th, 1894, meant to him—to me [meaning Sidgwick]—and to F. W. H. [meaning Myers]." The message went on: "I do not think they will find it hard to recall, but if so—let them ask Nora [meaning Mrs. Sidgwick]."

On March 14, 1906, in Mrs. Holland's automatic writing appeared the words "Eighteen, fifteen, four, five, fourteen, Fourteen, fifteen, five, twelve," and then the instruction to see the central eight words of *Revelations* 13:18.

On March 28, 1906, Mrs. Holland wrote (among other things) the words "Roden Noel," "Cornwall," "Patterson," and "do you remember the velvet jacket."

None of these things meant anything to the two automatists, but Alice Johnson eventually put it all together. The poem beginning with "Tintagel" in Mrs. Verrall's script was reminiscent of a poem by Roden Noel entitled "Tintadgel," a poem with which Mrs. Verrall was completely unfamiliar. The date of May 26, 1894, given in Mrs. Holland's script, was the date of Roden Noel's death. Although she had heard of Roden Noel and had read a couple of his poems (not, however, "Tintadgel"), she did not know him personally and did not know when he had died. But Roden Noel was known to both Myers and A. W. Verrall, and he was an intimate friend of Sidgwick, the purported communicator of this message, and Sidgwick's wife Nora. Dr. Verrall and Mrs. Sidgwick recognized this date as being the date of his death. Mrs. Holland herself had not attempted to look up *Revelations* 13:18, but Alice Johnson did and found that the central eight words were "for it is the number of a man." Taking this hint, she translated the numbers given in Mrs. Holland's script into letters (e.g., the 18th letter of the alphabet is "R"), and they thus spelled "Roden Noel." Finally, Alice Johnson learned that "Cornwall" was the topic of several of Roden Noel's poems, A. J. Patterson was a mutual friend of Noel's and Sidgwick's from their undergraduate days at Cambridge, and Noel had frequently worn a velvet jacket. None of this information was known normally to Mrs. Holland.

There were other things written in connection with this case by the two automatists, and Alice Johnson discusses in detail the ways in which subliminal associations to Roden Noel might have been awakened in the minds of the two automatists, even though they had no contact with each other; but in sum it seems highly unlikely that any of the connections to Roden Noel's poetry, his date of death, or other details were known normally by the automatist who wrote them.

Nevertheless, this case illustrates the other of the two major weaknesses of the cross-correspondence cases (the first being the highly specialized nature of the material involved)—namely, the difficulty of ruling out normal sources of knowledge that the automatists may not have consciously remembered or even been aware of (Stevenson, 1983). In the Roden Noel case and in many other cross-correspondence cases, this explanation seems highly unlikely, but the possibility nonetheless remains and cannot usually be ruled out completely.

One way to alleviate this problem and bring cross-correspondence–like material under more control might be deliberately to ask communicators or controls to produce references to the same topics in the writings of other automatists. An example of this was a suggestion made in Boston by Richard Hodgson to Mrs. Piper's control, on January 28, 1902, to "try and make Helen [Mrs. Verrall's daughter and another automatist] see you holding a spear in your hand" (Verrall, 1906:215). The control at first thought Hodgson had said "sphere," and although Hodgson tried to correct this mistake, there still seemed to be some confusion on the part of Mrs. Piper's control about whether the word was "spear" or "sphere." Three days later in England (that is, on January 31, 1902), Mrs. Verrall (not her daughter) wrote automatically in Greek and Latin words that referred to both a "sphere" and a "spear." At Mrs. Piper's next sitting in Boston, on February 4, 1902, the control said (through Mrs. Piper's automatic writing) that he had been successful in making himself appear holding a "sphear"—the misspelling indicating that the control still was confused about whether it was to have been a "sphere" or a "spear." The cross-references were put together only after February 13 and February 18, when Hodgson sent Mrs. Verrall reports of the two relevant Piper sittings. There had of course in the meantime been no communication between Hodgson in Boston and Mrs. Verrall in England.

Experiments of this kind, in which a suggestion is made through one medium to produce certain specified information through other named mediums, would, if successful, be an important step toward bringing cross-correspondences out of the realm of spontaneous material that we simply have to wait to receive and into the arena of experimental material that we can obtain under appropriate conditions of control. Nevertheless, although they might add valuable evidence for some kind of supernormal process occurring in mediumship, they probably would *not* take us much further toward saying that the supernormal process involves the survival of a deceased person. In the original cross-correspondence cases, the meaningful idea behind the various cross-references by the mediums apparently originated in the mind of a deceased person, and was in the mind of no living person. In these experimental suggestions for a cross-reference, in contrast, the idea would originate in the mind of the living experimenter, and therefore the hypothesis of telepathy from experimenter to mediums could always be invoked.

Drop-In Communicators

Let us turn next to the cases christened by Ian Stevenson (Stevenson, 1970) as "drop-in communicators." In such cases, a communicator completely unknown to any sitter present appears spontaneously in the sitting. Like the cross-correspondence cases, the motivation seems to have come from the communicator and not from any living person, since by definition no one at the sitting knows or recognizes the communicator. In such cases, it is only later (and sometimes many years later; see, e.g., Gauld, 1971) that investigators verify the accuracy of what the communicator said. And in the most important of these cases, the verified information contained in the communication was not obtainable through any single source, whether the memory of a living person or a document such as an obituary. Drop-in cases are thus considered important because, first, there is no motivation among the sitters to receive the communication, and, second, the information given was not obtainable from a single source, thus lessening the possibility that the medium had learned the information normally, but had forgotten (or falsely denied) having done so.

Drop-in communicators have been reported occasionally throughout the history of psychical research; perhaps the earliest were two cases reported by the SPR, one in 1890 and one in 1900 (see Myers, 1903(ii):471–477). As an example, I will summarize briefly a case that occurred in 1941, was first published in Iceland in 1946, and was finally published in the English literature in 1975 (Haraldsson & Stevenson, 1975a). At a sitting in Reykjavik, Iceland, on January 25, 1941, with the medium Hafsteinn Björnsson, at which three sitters were present, a communicator appeared who called himself Gudmundur or Gudni Magnusson. He said that he had been a truck driver and was driving over a mountain pass when his vehicle broke down. He had crawled underneath it, and had then ruptured something in his body. He managed to get home, but died while being transported across fjords by boat to medical care. He said that he and his death were connected with two towns, Eskifjordur and Reydarfjordur. He also said that his parents were living. The medium's control, called Finna, described Gudni as being a young man with blond hair that was thinning on top. None of the sitters recognized this person, but when one of them described the sitting two days later to a friend, the friend said that a cousin of his was married to a doctor in Eskifjordur. He wrote to this cousin, and subsequently nearly all the details given were found to be true, with some minor variations in detail, for a 24-year-old man named Gudni Magnusson who had died four months earlier in circumstances much like those described in the sitting.

There are quite a few drop-in cases just as impressive as this one—one such being that of Runki (or Runolfur Runolfsson), in which statements by the medium (again, Hafsteinn Björnsson) led not only to the identification of a man who had drowned 58 years earlier, but also to the discovery of a femur, presumably Runki's, buried in the wall of a house (Haraldsson & Stevenson, 1975b). Other cases include several occurring in a private circle that were investigated and reported by Gauld (1971). The Gudni Magnusson case, however, and the Gauld series are particularly important because there was a written record of the information given during the sittings, made before there was any verification of the case. In the Gauld series, detailed notes made at the time of the sittings had been preserved. In the Gudni case, several of the crucial details had been given in the letter sent to Eskifjordur to try to identify Gudni. Also, although there had been a newspaper obituary of Gudni, many of the important details given by the medium were not included in it, and there was no single source for all the verifiable information, other than the minds of Gudni's living parents. More drop-in cases of this quality would add greatly to the evidence for survival since the motivation behind the cases seems to come entirely from the deceased persons.[2]

Drop-in cases, however, suffer from two weaknesses. First, they are the easiest to fake. There are various means by which one can assess the likelihood that a particular case was faked, and in the best ones fraud seems highly unlikely. Nevertheless, Ian Stevenson and John Beloff (Stevenson & Beloff, 1980) investigated a medium who claimed to have produced numerous drop-in communicators, and they concluded that she was doing so fraudulently. The other major weakness of drop-in cases is that by definition they occur only spontaneously; we must wait for them to appear if and when they will. When they do appear, they are extremely important, but this is not a terribly efficient way to obtain evidence.

Proxy Sittings

I turn finally to the category of cases that I think are the most important to pursue in mediumship research today. Proxy sittings, as the name suggests, are sittings at which the person desiring a communication from a deceased loved one is not physically present at the sitting; instead a third person, preferably someone with little or no knowledge about the deceased person, arranges the sitting with the medium and attends it as a proxy for the real sitter. Proxy sittings were originally conducted as part of the effort of researchers to get beyond the theoretical super-psi/survival impasse. The reasoning was that, because there was no normal contact between the sitter and the medium, and because the sitter often did not even know whether or when a sitting would take place, the likelihood of a supernormal or telepathic interaction between the sitter and the medium was, if not eliminated, at least greatly reduced. At the time that most proxy sittings were conducted, in the 1920s and 1930s, this seemed a reasonable supposition. Unfortunately, as I mentioned earlier, experimental studies of psi have since shown that even extremely complicated tasks can be carried out. The important factor seems to be the goal rather than the complexity of the task. The super-psi hypothesis was greatly challenged by proxy sittings, but it was not by any means put to rest.

As I said at the beginning of this paper, however, I do not want to dwell on these theoretical issues. There are, in fact, other reasons why proxy research is important, reasons that are perhaps even more urgent now than they were in the 1920s and 1930s, since we have had so little good-quality mediumistic material over the past several decades. Proxy sittings are important first of all because they eliminate one of the major normal explanations, which is that the medium obtains or infers information from the sitter, such as by the sitter's verbal or behavioral responses to statements or by the sitter's appearance. In a good proxy sitting, that is, one in which the proxy has little or no information about the sitter or the deceased person, the proxy cannot provide any feedback to the medium; what one gets are the medium's unadulterated impressions or imagery. "Cold reading" is eliminated.

Proxy sittings also provide a means of addressing the other major normal explanation for mediumistic statements, which is that most of the statements are general or vague enough that they can apply to many people or be interpreted in a variety of ways by different sitters. On this hypothesis, apparently successful mediumistic sittings are simply the result of chance, arbitrary selection of statements, and overinterpretation on the part of a sitter biased by grief and wishful thinking. There have been some notable attempts to assess the likelihood of this explanation by developing quantitative methods of evaluating mediumistic material, and these efforts have resulted in a large literature in parapsychology on methods for evaluating free-response material in general, not only in mediumship research but also in other free-response research such as Ganzfeld studies, remote viewing, or dream telepathy (see, e.g., Burdick & Kelly, 1977; Schouten, 1993, 1994). I will not try to review this literature here, but will briefly mention some of the methods introduced in mediumship research in particular.

Quantitative Methods of Evaluating Proxy and Other Mediumistic Material

The first attempt to evaluate a mediumistic sitting quantitatively was made by James Hyslop in 1919, when he introduced the idea of giving a list of statements made by a medium at a sitting not only to the intended sitter, but also to "control" persons of comparable age, gender, and educational and social background, to see whether the controls would find just as much of the sitting to be accurate for them as the target sitter did. This first attempt was a massive effort: Hyslop listed 105 statements from a sitting that he had had with Mrs. Piper and sent them to 1,500 people, of whom 420 completed and returned the questionnaire. Hyslop was looking simply at the question of whether these 420 people would find overall that the 105 statements fit them as well as they fit Hyslop himself, and he found that they did not (Hyslop, 1919).

Similar studies in which controls evaluated sittings intended for someone else were conducted by Saltmarsh (1929), J. F. Thomas (1937), and Stevenson (1968), although no one has again used quite as many controls as Hyslop did. Stevenson, for example, made a list of 79 statements by a medium at a sitting intended for him, and gave this list to five male colleagues of approximately the same age, background, and profession as he. One of these, in fact, was his brother. Whereas 72% of the statements were accurate for Stevenson, only 46% were accurate for his brother, and only 30% were accurate for the other four controls. Stevenson introduced an interesting innovation, however, by classifying the statements on three dimensions: first, whether they were objective (that is, independently verifiable) or subjective (that is, requiring someone's judgment); second, whether they were general or specific in nature; and third, whether they were matters of public knowledge or intimate personal details not in the public domain. To evaluate the criticism that sitters score statements more generously because they know they are intended for them, whereas controls score more stringently because they know they are *not* intended for them, Stevenson compared the objective statements, which required no judgment and could only be scored as accurate or not, with the subjective statements, which could be subject to biased responses. For him, 70% of the objective statements were accurate, for his brother 55% were accurate, but for the other four controls only 33% were accurate.

Before Stevenson, there had also been other attempts to assess the probability of a statement being accurate, that is whether it was so general it could apply to many people or, conversely, so specific it would apply to very few. Saltmarsh (Saltmarsh & Soal, 1930) was the first to attempt this, but his method was highly dependent on the subjective opinion of the judges assigning probability values to statements. Pratt therefore developed a method, in his research with Mrs. Garrett (Pratt, 1936, Pratt & Birge, 1948), to make the probability evaluations less subjective: He had a large number of people rate the accuracy of individual statements for themselves, and then applied a complicated statistical formula to determine the probability of a statement's accuracy among the general population. Perhaps the most important aspect of Pratt's method, however, was that he was the first to use all proxy sittings so that none of the people scoring statements had any idea which were intended for them and which were not. In previous studies in which controls were asked to evaluate statements, the controls knew the sittings were not intended for them. Although one can assume that in most of these studies the participants were motivated to evaluate the statements as objectively as possible, and although Stevenson (1968), as mentioned above, introduced an important way of assessing the objectivity of the evaluations, the basic principle introduced here by Pratt of giving sitters multiple readings for blind evaluation is an important procedural innovation, providing a simpler and more satisfactory way of evaluating the chance hypothesis.

There are two basic methods for evaluating free-response material: One can score the material item by item and calculate the combined accuracy of all the statements together. Pratt and the few others who have also tried to evaluate proxy sittings quantitatively (e.g., MacRobert, 1954, Schmeidler, 1958) used this item-based procedure. A second way of evaluating material is to rank the material globally, that is, by considering the material *overall* rather than statement by statement. West (1949) introduced this method into the quantitative evaluation of mediumistic material.

In my view this global method of rating or ranking entire readings should be adopted more widely for any future research with mediums, as it has been, for example, with Ganzfeld studies (e.g., Bem & Honorton, 1994). First of all, evaluating readings on an item-by-item basis requires highly complex and time-consuming methods both of scoring and of statistical evaluation. Moreover, many statements are not independent of others, which further complicates the scoring and evaluation. The global method is a much simpler and more straightforward way of determining whether a medium's accurate statements apply specifically to the intended sitter or whether they are general and vague enough to apply to many people. Second, and perhaps more importantly, this global evaluation allows for the likelihood that much of what a medium says in a sitting is in fact what might be called "filler" material. Just as in a Ganzfeld session, much of the medium's imagery and impressions may come from his or her own mind and have nothing to do with the intended target; but if, on the basis of a few important and highly specific statements, sitters can pick out their own readings from a group more often than one would expect by chance, then we have grounds for attributing this part of the medium's statements to some supernormal process.

Global evaluations like this, however, can be carried out only when sitters do not know which was their own sitting—that is, with proxy sittings. In the remainder of this paper I will give an overview of the history of proxy research, together with a few examples of actual proxy sittings, primarily to show just what is possible and the kind of results we might aim for in proxy-sitting research. A general point I want to make in starting off, however, is that, despite the potential importance of proxy sittings for eliminating the common normal explanations of successful mediumship, very little such research has been carried out and reported; the vast majority of what has been done took place in the 1920s and 1930s, and even then it was limited primarily to two mediums, Mrs. Osborne Leonard and Mrs. Warren Elliott, and five researchers.

Research by Nea Walker

The first report of a proxy case was made by Nea Walker, who was Sir Oliver Lodge's assistant in psychical research. Many bereaved people wrote to Lodge because of the books he published about his experiences with mediums, and Nea Walker handled most of this correspondence. One letter was from a Mrs. White, a woman who had lost her husband in 1920. Her initial letter led to a long series of proxy sittings carried out by Nea Walker on behalf of Mrs. White over a period of three years, with Mrs. Leonard primarily, but also with Miss Walker's sister, Damaris, who was an amateur, or private, automatist. Nea Walker published a detailed report of this case in 1927 in a book called *The Bridge* (Walker, 1927), which included not only full reports of the proxy sittings, but also reports of some anonymous sittings that Mrs. White herself had had with Mrs. Leonard and other mediums. Because the case involves so many sittings, with several mediums, and under various conditions, it is not easy to summarize; my overall impression is that many specific details were given, not only in regular sittings but more importantly in proxy sittings, and that Nea Walker was meticulous not only in controlling the conditions, but also in describing them.

The next report of a proxy case was also by Nea Walker (Walker, 1929). This was the Tony Burman case, and it was much less complex because there were only three proxy sittings by Nea Walker, one with Mrs. Garrett and two with Mrs. Leonard. Tony Burman had died in a motorcycle accident in 1926. On the day of his accident, but several hours before his mother learned about it, she was having a sitting with Mrs. Garrett. During the sitting, and about two hours after the accident occurred (but before Tony had actually died), Mrs. Garrett's control Uvani told Mrs. Burman that she saw her "littlest son," who was saying that there was "some trouble." A year later Nea Walker held the three proxy sittings for Mrs. Burman. The one with Mrs. Garrett and one of the two with Mrs. Leonard were successful. For example, among other things, Mrs. Garrett, who had no reason whatever to connect Nea Walker with Mrs. Burman, said that the communicator had tried to warn his mother a year earlier that "all was not well," this apparently being a reference to the sitting Mrs. Burman had had with Mrs. Garrett on the day of her son's accident. Also, Mrs. Leonard was nearly able to get the family name, in the groping manner that is so common in mediumship. She began by saying it was a "B" name, then said Borrowman, and then later in the sitting she returned to this by saying Burry, Birry, Bur, not Bro, but Bur, Burnam. The proxy, Nea Walker, knew this name; but other details that came out during the sitting seemed to be about matters unknown to her. For example, the communicator mentioned an argument about a hat; he had had this argument with his father on the day of his death, although no one but his father knew this. There were also some comments about gloves that he should have worn. His not wearing rubber gloves at work had led to a dangerous poisoning of his hand, and his family had made a private "appeal" to the deceased Tony before the sitting that this incident be referred to.

Nea Walker made a third major contribution to proxy research in 1935, when she published another book, *Through A Stranger's Hands* (Walker, 1935). One important criticism about her first case, the White case, was that over the five years that she worked with Mrs. White, she had become closely involved emotionally with the case and had eventually learned a great deal about Mr. and Mrs. White.[3] In her second book, therefore, she addressed the question of whether she could have successful proxy sittings for people who remained distant to her, and this book is a detailed report of eight such cases involving sittings conducted from 1929 to 1933. The medium for all was again Mrs. Leonard. For this study Miss Walker chose, from among the many people who wrote to her and to Lodge, first, people who had included very little information about the deceased person in their letters, and second, people who seemed educated and intelligent since, as she pointed out, much depends upon the ability of the person annotating sittings to read them carefully and with an open mind. She also tried as much as possible to vary the age, gender, and mode of death of the deceased person. For this series, no more than three proxy sittings were held for any one case, and the families were not told when, or even whether, a sitting would be held.

There were interesting, specific, and apparently veridical details in all eight cases, although there was also much that was vague and not specific. I would like to give some excerpts from one of the cases, to give an idea of what they are like, the vague statements as well as the specific ones; but I first want to make one essential point, that these are *excerpts* only. One of the most important features, not only of this book, but also of all the mediumship research of the late 19th and early 20th centuries, is that verbatim shorthand records were made during the sittings.[4] Moreover, in many reports of proxy cases from the 1920s and 1930s, virtually the entire transcript of the sitting was published; whenever material was omitted (which was often necessary since sittings with Mrs. Leonard could run for a couple of hours or more), the reporter would give a description of what was omitted and why, usually because it was not related to the particular case in question or because it was non-evidential matter such as remarks about problems of communicating, conditions in the afterlife, and so forth. Publishing entire, or near-entire, transcripts is important in mediumship research so that readers can see for themselves exactly what was said, when it was said, and in what context, both by the medium and by the sitters. For example, as I described earlier, in the Tony Burman case the medium mentioned a name that was very similar to "Burman", and we can hope that Nea Walker, who knew this name, as an experienced sitter would have given no hint to the medium that she was on the right track in her attempt to get the name. But, without a complete transcript (and probably also a video recording of the proxy sitter), we cannot be certain that some such hint did not occur. This is of course the reason for keeping the proxy's knowledge to a minimum; but even then having complete transcripts is important for other reasons. First of all, some statements are important or evidential only in their entire context, and may lose their meaning—or, conversely, take on undue meaning—when taken out of context. More importantly, perhaps, it is only with entire transcripts available that we can begin to analyze and understand a medium's habitual thought processes, trains of association, expressions, habits, symbolic images, and so forth. Saltmarsh, for example, had noted that Mrs. Elliott had "an unconscious predilection for certain names and initials," although he also noted that the frequency of her use of these names did not correspond with their frequency in the general population (Saltmarsh, 1929:97). Nevertheless, if, for example, we find that a medium uses the unusual name Jedediah in every other sitting, then its evidential value when it does finally correspond to a real person is considerably lessened.

With this cautionary statement about drawing conclusions from excerpts alone, I will now give a few excerpts from one case in Nea Walker's second book (Walker, 1929). In November 1932, Lodge received a letter from a Dr. van Tricht, who six months earlier had lost his two children, a boy age 10 and a daughter age 2½. He and his family had been on a ship, returning from the Dutch East Indies, where he practiced medicine, for a holiday back in Holland. On the ship, the parents were in one room, the children and their governess in another. Fire broke out on the ship one night, and the parents were able to escape, but the children and the governess were not. All of these details were in Dr. van Tricht's letter, and as Miss Walker herself said: "I would rather have had less information, but that could not be helped" (Walker, 1929:381). Nevertheless, an important point to remember with regard to proxy cases is that, even when the proxy has had some degree of knowledge about the sitters or deceased persons, and could have provided clues to the medium in some normal way, much important information *not* known to the proxy has also been given.

Almost immediately in the first sitting intended for Dr. van Tricht, Mrs. Leonard's control Feda said that there was a boy here, a boy who went over quickly, and (later in the sitting) that he had not been an invalid but healthy. She said the initial R had something to do with him (his name was Rudolf, a fact known to Nea Walker), and then she said: "The day before he passed over I have a feeling that there was something happened, arranged, only the day before, that he won't say had made him pass, but had led to him passing" (Walker, 1935:387). Later in the sitting Feda said:

> *Again* I got "the day *before*". Something was said, and done, and arranged, the day before. And, Mrs. Nea, if something had been done, acted on, that was suggested the day before, this passing might not have happened. This is true, though I don't know if it's good to say it. There was a suggestion made, . . ., the day before, that might have altered everything. . . . Anyhow, there are some things you can't alter. It looked, afterwards, as if—"If only we had done what we thought of the previous afternoon, all would have been altered." (Walker, 1935:393–394)

And then still later in the sitting: "And they remembered about altering rooms, and one taking one, and one taking the other. Changing rooms was just before. People going from one room to another—'You will have to take that other one'" (Walker, 1935:400). Nea Walker had no idea what all this referred to, but the mother later explained that the day before the fire she and the governess had discussed, in the presence of the 10-year-old boy, changing the room that the children and governess were in, but they had not yet gotten around to doing this when the fire occurred.

Feda also said "I don't know if you would know if someone rather young— I can't tell the sex—passed near the same time as himself. I am getting a second . . . it's someone with him. Also very young. And there's a link between them. They are together. Two young ones together" (Walker, 1935:391). Later she said: "And were there five of them in a group that had been all together? Five of them. They have been used to five people, all together, a group," apparently a reference to the parents, children, and governess (Walker, 1935:394).

There were also several attempts at names. For example, at one point during the sitting Feda said: "I keep getting initials A. W. . . . It's someone who's passed over who's looking after them over there. A. W." (Walker, 1935:406); and then later she said: "And will you say Alfred has helped them" (Walker, 1935:412). At first the parents couldn't place "Alfred," but they later remembered that the family had befriended a man on ship whose name was Alfred L. —W. (the full name was not given by Miss Walker, nor had it been known to her before the sittings). He had survived the fire, but had been killed a few days later in a plane crash.

There were many other interesting details in this case, but I want to give one final example which illustrates the importance of having a verbatim transcript and of looking at a case in its entire context. During this sitting, Feda said:

Do you know Bob, or Bobby, in connexion with him? . . . Not passed over. Someone that he would have been with not long before he went over. . . . A man . . . a man who was very much with him, very fond of him, but not related to him. An older man, that this boy was very fond of on the earth. Not his father. Not related at all, and yet he was awfully fond of him, looking upon him, and up to him, almost as you would a relation. Petter, Payter, Petter, Pitter, Peter, Pellets. . . . Now I am getting a name sounding a bit like Peter or Peters . . . Peter and a place on the earth that it was very much to do with. (Walker, 1935:388–389)

During a second proxy sitting three weeks later (the transcript of which, unfortunately, was not printed), Feda again referred to an elderly man, Bobby, and Peter in succession, and after referring to the deceased boy's Christmas holidays a year earlier, kept giving the name "The Mount." Taking all these items together seemed to suggest a very close family friend with whom the family often visited, a man the children referred to as Uncle Bottie. Uncle Bottie's last name was Diemont (pronounced Dee-Mont), and Peter was the name of the boy the children often played with when visiting the Diemonts (Walker, 1935:389). Although "Bobby", "Peter", and "The Mount" looked at in isolation are not terribly impressive items, since they are common names, the clear and close association between them and the way in which they are presented in the entire context make them much more impressive.

Research by H. F. Saltmarsh

An even more sustained body of research with proxy sittings was reported in a 1929 paper by Saltmarsh. The medium Mrs. Elliott agreed to give three sittings a week for a year for this project. Eighty-nine of the 142 that she eventually gave were proxy sittings, and the rest were anonymous sittings with the real sitter present. The proxy sittings were extremely well-controlled, in that sitters sent an object belonging to the deceased person to the SPR, and it was labeled by code and stored securely by someone at the SPR. On the morning of a sitting, one object was selected at random in such a way that the sender of the object did not know that his or her object had been chosen, the proxy did not know whose object had been chosen (and in any case knew nothing about the pool of sitters and objects), and the person at the SPR who received and numbered the objects and kept the records did not know which object had been chosen on any particular day.

Although the research was well-controlled in keeping the proxy sitters completely blind, it was not as methodologically sound in the evaluation phase. One purpose of this research was to examine the hypothesis that good sittings are simply the result of "commonplace or vague statements which would be probably true of most sitters" (Saltmarsh, 1929:58), and to this end Saltmarsh had the real sitters annotate their sitting, as well as a number of pseudo-sitters, or controls, who otherwise had no connection with the research. Unfortunately, the sitters all knew whether the readings were, or were *not*, intended for them. Saltmarsh tried to lessen any effect of bias in the scoring based on this knowledge by stipulating that the number of correct items had to be at least eight times higher for the real sitter than for the controls, a choice Saltmarsh fully acknowledged as being "to some extent arbitrary" (Saltmarsh, 1929:50). Saltmarsh himself assigned an overall score to each of these annotated records by a system in which he categorized statements as either (1) vague, (2) definite but true of many people, and (3) highly specific. Again, he recognized the subjectivity of this system, but he emphasized that, since he had done all the scoring, although the scores could not be viewed as absolute in any sense, they could be valuable for comparative purposes.

Perhaps some of the most interesting findings came from a series of 19 randomly chosen proxy sittings that were annotated by an average of four to five controls each. Taking the 19 sittings as a whole, the real sitters' scores were more than 12 times higher than the control annotators' scores. For example, in one case the real sitter had a score of 59, whereas among the five controls, one had a score of 3 and the other four had scores of 0. In another case, there were nine controls, and the real sitter had a score of 62, and control scores were 14, 11, and 7 0s. Again, it is unfortunate that Saltmarsh did not use blind scoring in this study; but the large differences in the scores of real and control sitters are at least suggestive that something more than bias in annotating was at work here.

Saltmarsh made a number of other interesting observations in this study that might ultimately prove useful in theorizing about the source of information given by mediums. First, his overall results showed that sittings at which the sitter was present got higher scores than the proxy sittings did. Nevertheless, as the above-mentioned analysis of proxy sittings showed, many of the proxy sittings were highly successful, even if at a lower level of scoring. Another interesting finding was that when Saltmarsh classified statements as to whether they referred to premortem information or to postmortem events, there were about an equal number of statements in both groups, but the veridicality of the statements was actually somewhat higher in the postmortem group, indicating (depending on one's interpretation) either that medium had become aware of postmortem events by telepathy or that the deceased person remained aware of postmortem events concerning his or her loved ones.

Research by C. Drayton Thomas

Another person who contributed importantly to proxy research was C. Drayton Thomas. In a 1932 paper, he discussed a series of 24 cases of proxy sittings that he had held with Mrs. Leonard, beginning as far back as 1917. In most of these cases there were only one or two sittings, but in one there were four and in another 11. The cases also involved a variety of conditions: In half of the cases, Thomas had met and knew something about the deceased person or the family, although usually not much. The educational and social background of the people concerned varied widely, and the deceased persons ranged from young children to elderly people. Finally, as Thomas commented, there was "a rather bewildering variety of results" (Thomas, 1932–1933:139), of which he tried to make some sense in his paper. After having the person asking for the proxy sitting annotate the sitting, Thomas and three other members of the SPR (including Saltmarsh) assigned each sitting a value as to its evidential quality, and he then divided the cases into four groups: four cases that they considered good, four cases that they considered fair, seven that they considered poor, and nine that they considered either inconclusive or outright failures. In the paper, Thomas gave brief summaries of two cases from each of the four categories to illustrate their nature, and he then provided in an appendix a more detailed report of a ninth case, which the evaluators considered the best one.

The main purpose of the study had been to evaluate the major interpretations, telepathy or survival; but, as Thomas said, the key question running through his analyses was "Why should the results differ so widely?" (Thomas, 1932–1933:150). He considered several factors in addressing this question. For example, the involvement of the families seemed to make no difference: There were failures when the family was highly motivated and thinking strongly about the deceased person at the time of the sitting, and successes when the family did not even know that a sitting was taking place. Strong emotion is often thought to facilitate psi, but in the case in which there seemed to be the strongest emotion and desire to communicate on the part of the family, the sitting was a failure (Thomas, 1932–1933:151). Thomas noted that this was also the only case among the 24 in which the deceased person had died some years earlier, and he wondered whether elapsed time since death is a factor. There was also no difference in results among the 12 cases in which Thomas had some acquaintance with the family, as compared with those 12 in which he did not (Thomas, 1932–1933:159).

Thomas did, however, note some differences that might serve as hypotheses for future research. He noted a tendency for cases to be better when the deceased person was a young adult rather than a child, a middle-aged adult, or an elderly person. He also thought, on the basis of the information available to him, that in the successful cases the deceased person had been an educated and intelligent person; but he went further and suggested that it was not so much native intelligence alone that was a factor, but the possession by the deceased person of an alert, interested, and active mind (Thomas, 1932–1933:162–163).

In general, Thomas's conclusion was that the success of sittings depends primarily on characteristics associated with the deceased person. As he put it:

> To those who are prepared to admit the possibility of human survival and communication, I ask, Is it not natural that some should have greater aptitude than others for the difficult and delicate operation of transmitting their thoughts through an intermediary, and of making suitable selection of evidential matter? We know how widely aptitude for selection and expression of ideas is found to vary in mankind. Some can select with finer judgment and can express themselves with greater precision; this being so on earth, one is not surprised to find indications of it in communications from the discarnate. And further, few of us can have failed to notice how widely people differ in their regard for relatives. Not all feel the same urge to set at rest the minds of friends who may be anxious about their welfare or desirous of hearing from them. Such differences may quite naturally persist in the life after death, some being very desirous of communicating, others much less so.
>
> In this diversity of mental ability I find a cause for the wide difference shown in my series of proxy sittings, a difference ranging from complete failures to clear-cut success. (Thomas, 1932–1933:167)

In support of his overall conclusion that something about the deceased person, and not so much about the families or proxies, is the important variable, Thomas noted that, in his experience with Mrs. Leonard, communicators who failed at a first sitting were never successful when a second attempt was made, whereas those who were successful at a first sitting were often just as successful at subsequent ones (Thomas, 1932–1933:166–167).

After this paper on a series of proxy cases, Thomas went on to publish many other papers and books on various aspects of his work with Mrs. Leonard, but I will deal here only with three additional proxy cases that he published as individual case reports. The first of these cases, the Bobbie Newlove case (Thomas, 1935), is one of the most important cases in the history of mediumship in general. In September 1932, Thomas received a letter from a complete stranger, a Mr. Hatch, who lived about 200 miles away, saying that his 10-year-old step-grandson, Bobbie Newlove, who had lived with Mr. Hatch his whole life and was like a son to him, had died suddenly of diphtheria the month before. From November until the following June, Thomas had 11 proxy sittings with Mrs. Leonard for Mr. Hatch, at first without Mr. Hatch's knowing he was doing so and later at times unannounced to Mr. Hatch or his family. Numerous quite good details, about which Thomas knew nothing, were given in the course of these sittings, such as a description of a photograph in which Bobbie was wearing an unusual costume (Thomas, 1935:452), a remark about hurting his nose shortly before he died (Thomas, 1935:452), and detailed descriptions of his village and the location of his house in it, including the name Bentley, which was the name of the street where his school was located. He mentioned that there was a broken stile on the way to some place he used to go. His family knew nothing about this, but they later learned that there had been such a broken stile on a path toward a place Bobbie often played, although the stile had been taken down shortly before his death (Thomas, 1935:455). At another sitting he described a place near his home, saying "There is a place 'C'—close by, a long name sounding like Catelnow, Castlenow. There seemed to be two or three syllables, like a Ca sound, cattle or castle something" (Thomas, 1935:478). Mr. Hatch wrote that the last place Bobbie went, the day before he became ill, was Catlow, a small village near their town.

By far the most important feature of the case, however, is what Thomas called "The Problem of the Pipes." In several of the 11 sittings, the communicator built up a picture of what had probably caused Bobbie's death, giving facts and details that were completely unknown to his family until, after the 11 sittings were complete, Thomas visited the family on two occasions and together they verified what the communicator had been saying. Feda began this in the second sitting by saying that it was not just the diphtheria that had killed him, but something else that affected his heart and weakened his system (Thomas, 1935:483–484). In the third sitting, she elaborated by saying something had happened nine weeks before his death that was a link to his passing, and then she added: "Wait a bit, 'pipes, pipes'; well, he says just this—'pipes'. That word should be sufficient. Leave it like that" (Thomas, 1935:484). The family had no idea what this referred to. Feda returned to the topic in the fourth sitting, saying there was something that had made it easier for him to get sick and also then to be unable to shake it off. She then added: "I don't know what you mean, Bobbie, you say you got yours from the pipes" (Thomas, 1935:485). In an effort to help the family understand this, Thomas asked Feda for more information in the fifth sitting. She said it was not at his home, but at a place he went to with which his family was not familiar; he described the place and said that another boy went there with him. Feda again said "through these—what he calls the pipes—he picked up the condition which was not the cause of the trouble in the first place, but it introduced a destructive element which resulted in diphtheria." She then added: "Either before or after Bobbie caught it there—we think after—there was something done to apparently improve matters with regard to those 'pipes'. There was something altered that probably now has improved the condition, made it safer" (Thomas, 1935:486). In the sixth sitting, Thomas again asked for more help with identifying and locating the pipes. Feda gave the description of Bobbie's town and of a route going somewhat out of town (Thomas, 1935:490–492). Finally, in the tenth sitting, there was a more detailed description of the place associated with the pipes and how to get there (Thomas, 1935:494–499).

In June 1933, Thomas visited the family and read a diary that Bobby had kept, in which he said that on June 15, a date nine weeks prior to his death, he had been with his "gang," which his family learned was a secret society consisting of himself and a friend, Jack. They also learned that the boys had often gone to play at a place outside the town called The Heights (Thomas, 1935:485, 488). At a second visit the next month, in July, Thomas and Mr. Hatch, on the basis of the information that had been given in the sittings, located the place where Bobbie and Jack, unbeknownst to the family, had played, and also a pipe from which groundwater fed some ponds there. In September, Mr. Hatch discovered a second pipe there (Thomas, 1935:488), confirming Feda's use of the plural word "pipes." In November 1933, Mr. Hatch wrote to Thomas to say that the boy Jack had confirmed that he and Bobbie had played with the water at the pipes (Thomas, 1935:488); and in February 1934, Mr. Hatch received a letter from the local health officer confirming that the pools into which the pipes poured water were liable to contamination that could cause illness, although the water issuing from the pipes themselves was safe (Thomas, 1935:501).

As Thomas points out, there was no direct confirmation for the opinion of the communicators and, apparently, of Bobbie himself that these pipes had been a factor in causing his death (Thomas, 1935:501). Nevertheless, as Thomas also points out, this lack of direct confirmation is inconsequential, since the facts on which this opinion was based were given in the communications, were completely unknown to the family, and were verified by information given in the sittings. Most importantly, the only person in possession of all the information concerning the pipes and their possible connection to Bobbie's last illness was Bobbie himself. As Thomas put it, "Anyone acquainted with these facts might have suspected that the throat infection . . . was traceable to the contaminated water. But no one on earth had the least suspicion of this until it was stated in the course of these sittings" (Thomas, 1935:501–502).[5]

The next case is also a fairly well-known one (Thomas, 1938–1939). It is particularly interesting, not only because of some quite specific details that were given, but also because Thomas, as the proxy, had no direct contact with the deceased person's family; Professor E. R. Dodds served as an intermediary between the family and Thomas, again in the hopes that this additional barrier between the medium and the family would weaken the hypothesis of telepathy. The deceased person in this case, Frederic William Macaulay, had died in 1933, and three years later Dodds wrote to Drayton Thomas suggesting that he try to contact this person at some proxy sittings. A first attempt with another medium was unsuccessful, but beginning in June 1936 and continuing for a year, eight proxy sittings relevant to this case were held with Mrs. Leonard. Unfortunately in my view, rather than print the transcripts more or less as a whole, in succession, Thomas in this paper chose to group excerpts from the sittings related to three main topics: items relating to the communicator's professional life, items relating to personal, intimate memories, and items relating to friends and acquaintances. Nevertheless, it is clear from these excerpts that some specific details emerged, details of course unknown to Thomas or Mrs. Leonard. I will briefly describe three of these that seem particularly good because they are quite personal, yet also quite specific. At the third sitting, Feda said: "There is also a John and a Harry, both with him. And Race . . . Rice . . . Riss . . . it might be Reece but sounds like Riss" (Thomas, 1938–1939:265). The annotator, Mr. Macaulay's daughter, explained:

> The most interesting passage is "It might be Reece but it sounds like Riss". This carries me back to a family joke of these pre-war days. My elder brother was at school at Shrewsbury and there conceived a kind of hero-worship for one of the "Tweaks" (sixth form boys) whose name was Rees. He wrote home about him several times and always drew attention to the fact that the name was spelt "Rees" and not "Reece". In the holidays my sister and I used to tease him by singing "Not Reece but Riss" until my father stopped us, explaining how sensitive a matter a young boy's hero-worship was. I think Rees was killed in the Great War. (Thomas, 1938–1939:265–266)

At the fourth sitting, Feda said: "I get a funny word now . . . would he be interested in . . . baths of some kind? Ah, he says I have got the right word, baths. He spells it B A T H S. His daughter will understand, he says. It is not something quite ordinary, but feels something special" (Thomas, 1938–1939:266). The daughter replied later:

> This is, to me, the most interesting thing that has yet emerged. Baths were always a matter of joke in our family—my father [who was a water engineer] being very emphatic that water must not be wasted by our having too big baths or by leaving the taps dripping. It is difficult to explain how intimate a detail this seems. A year or two before his death my father broadcast in the Midland Children's Hour on "Water Supply" and his five children were delighted to hear on the air the familiar admonitions about big, wasteful baths and dripping taps. (Thomas, 1938–1939:266)

A little later in the same sitting Feda said: "What is that? . . . Peggy . . . Peggy . . . Puggy . . . he is giving me a little name like Puggy or Peggy. Sounds like a special name, a little special nickname, and I think it is something his daughter would know. Poggy, Puggy or Peggy. I think there is a 'y' on it" (p. 269). The daughter replied: "My father sometimes called me 'pug-nose' or 'Puggy.'"

Items like these three, and especially unusual nicknames, are in my view the ideal type of information to aim for in sittings: items that are both highly unusual and so personal that they are unlikely to be known by anyone outside a few close family members.

Another interesting feature of the Macaulay case is that the family had had a few sittings with another medium, Mrs. Brittain, shortly after Mr. Macaulay had died, and Thomas outlines what seem to be references by both mediums, three years apart, to the same information (Thomas, 1938–1939:300–306). For example, Mrs. Brittain had said: "Now he will show something very important so that you will know it is really him. It is a black thing—a telescope" (Thomas, 1938–1939:302). Three years later, Mrs. Leonard said: "Does he hold something to his eye? Not an ordinary glass, but something he held up right close to his eye, like looking down a peep hole—sort of little tunnel" (Thomas, 1938–1939:303). The item in question, not at first recognized by his daughter, was a telescope used with a rifle. As the daughter explained, she later asked her mother whether her father had had a telescope:

> She looked at me as if I were mad, and then, for the first time, I remembered that my father was, until about five years before his death, a most enthusiastic rifle shot.... I can't understand this complete lapse of memory on my part. He used the telescope on the range. (Thomas, 1938–1939:303)

This example of an initial failure to recognize the accuracy of a statement raises another problem, that of the limitations and idiosyncracies of individual *evaluators* of sittings. One person may not have (or remember) all the necessary information. Moreover, although some people may be loose or lenient in their evaluations, others err in the opposite direction. As Saltmarsh (Saltmarsh, 1929:136) remarked, the idiosyncracies of annotators "cuts both ways. Some annotators are so ingenious in finding correspondences that their results require a heavy discount, others are so refractory that they will not see anything but the most direct hits." Other shortcomings of individual evaluators are even more difficult to account for. West (1949:100) gave an example of a woman who "said 'no' to the statement 'strong clerical associations' although her father is a clergyman and she has lived with him almost all her life." One solution to such idiosyncracies may be to have more than one person annotate individual sittings; as Saltmarsh (1929:137) said: "It would seem that the only possible method of eliminating this error is to multiply the number of cases, sitters, and annotators, to such a degree that the individual variations average out."

Another Drayton Thomas proxy case worth describing briefly is the Aitken case (Thomas, 1939). An unusual feature of this case is that there is also a drop-in aspect to it: At a sitting with Mrs. Leonard in 1928, the communicators asked whether Thomas had received a letter from a middle-aged man about his son. When Thomas said no, they explained that he soon would, that it was an accident case, connected with a motor car, and that the young man was killed outright or nearly so. They mentioned that the name Morton, or a like-sounding name, was involved, and they then said that the father who would write had once lived near a place where Thomas himself had once lived. Eleven days later Thomas received a letter from a Mr. Aitken, whose son had been killed outright, not in a motor car accident but in an Air Force accident 11 months earlier. Moreover, as Thomas later learned, Mr. Aitken had lived for 12 years in Norton, where his son was born and lived, and this was a town that Thomas had lived in for two years more than 30 years earlier.

After this initial "drop-in" appearance and the subsequent arrival of Mr. Aitken's letter, four sittings with Mrs. Leonard were held for Mr. Aitken, with Thomas as the proxy sitter, and some additional details unknown to Thomas were given. At one of these sittings, Feda said:

> There was somebody else he [the deceased son] was interested in, that perhaps you [his father] don't know . . . a name that starts with B, and I think there is an R in it . . . it's not a long name—very much linked with him . . . it might be a Mr. BRICK. . . . I feel this is something you could use for building. (Thomas, 1939:122)

At a sitting two weeks later, Feda again mentioned:

> a name starting with BR—rather an important name with him . . . somebody he was linked up with shortly before his passing. . . . I also want to know if there is anything to do with him like a little ship . . . or a little model of a ship. . . . He is showing me something like a toy ship—a fancy ship, not a plain one—'laborate, rather 'laborate—with a good deal of detail shown in it—it seemed to be connected with his earth life—but some time before he passed over, rather early in his earth life. (Thomas, 1939:122)

After the first proxy sitting, the deceased man's brother had, without telling anyone else, made a mental appeal to his deceased brother to send a message through the medium about a mutual friend of theirs in the Air Force who had recently been killed. The friend's name was Bridgen, their parents did not know him, and before joining the Air Force he had worked at a firm that made scale models of ships, photographs of which he had shown to the two brothers. The living brother told Thomas that he had expected that, if the medium was unable to get Bridgen's name correctly, she would get something about these model ships.

Another Drayton Thomas case that I want to mention briefly is also a very well-known case, the Edgar Vandy case (Broad, 1962:349–383, Gay, 1957, Mackenzie, 1971). This was not primarily a proxy case, although Thomas held two proxy sittings for Edgar Vandy's brothers, one shortly after his death and one about ten months later. What is primarily interesting about this case is that there were several mediums involved (including one sitting held 23 years after Edgar Vandy's death) and two primary sitters (Edgar Vandy's two brothers), and there seemed to be many cross-references to common topics by all the mediums. In addition, similar cross-references appeared in the two proxy sittings that Thomas held with Mrs. Leonard. Like the Aiken case, there was also a drop-in aspect to this case: One of the Vandy brothers wrote to Thomas shortly after his brother's death, saying only that he had recently lost a brother, that there was some doubt about how he had died, and that he would like Thomas to hold a proxy sitting for him. At Thomas's next sitting with Mrs. Leonard, although he had had no intention of having this be a proxy sitting for the Vandy brothers, there nevertheless appeared a communicator who seemed to be referring to topics that Thomas later learned dovetailed with topics that the two Vandy brothers had been hearing about from other mediums.

Research by J. F. Thomas and Lydia Allison

The next case I can again refer to only briefly here because it is a long and extremely complicated one, and this is the John F. Thomas case (this Thomas being no relation to Drayton Thomas). After John Thomas's wife died in 1926, he began a long series of sittings with many mediums, which he eventually reported in a Ph.D. dissertation under the direction of William McDougall at Duke, and then in a book (Thomas, 1937). There were ultimately 525 sittings with 22 mediums in Boston and in London, 352 of which were proxy sittings, and the rest of which were sittings that Thomas or his son attended, nearly all of them anonymously. Eighteen of these sittings, including 16 proxy sittings, all with Mrs. Leonard, were published in Thomas's book in detail, with complete transcripts of some of them. Additionally, Lydia Allison, who held 21 proxy sittings for Thomas with four mediums in England, published transcripts of four of these proxy sittings that she had with Mrs. Leonard (Allison, 1934). Numerous quite specific and veridical details were given in the course of these many sittings, of which I can here give only one example. During Mrs. Allison's fourth proxy sitting with Mrs. Leonard, in June 1929, at one point she asked Feda where the deceased communicator and her husband had lived. (Although she had met John Thomas only once and knew little about him, Mrs. Allison did know that he lived in Detroit.) After some rather vague remarks, including "What do they pack in cases? . . . you can eat something that comes from there," Feda suddenly said:

> What do they cut? Cutting something. Chopping and cutting. This isn't to eat at all. Because I am getting the feeling of steel and metal, then sharpness—cutting. I do get a feeling of metal. I think that must have a good deal to do with metal. Is there some factories there? Because I feel noises. Because I—clank, clank, clank—and fitting. They fits things together, stamping and cutting out. More of fitting. Some isn't fitted like parts of them is made. I feel some rather big pieces and what's the circles, wheels, that I see? Because I see wheels and circles and all sorts of round and square things and hundreds and hundreds of men working. No, . . . thousands and thousands; and the whole place is like a beehive of humanity working in these huge places. . . . It is noisy—

it's hammering on metal—ringing noises. They runs along, they runs along. Like this. [Imitating whirring sounds.] . . . Some are sent out incomplete and some complete. Because some are assembled in other places. Wait a minute. . . . What did you say about "D"? . . . Detra—Detra. Something beginning with "D". Detra—Detro. He tried to say it. . . . Detra—he's trying to say the "D" word again. Detra—Detroi—it is important but I can't get it. (Allison, 1934:132–133)

As Lydia Allison herself remarked, "To an American it seems almost incredible that Detroit and the automobile industry should not be immediately associated" (Allison, 1934:134). Nevertheless, she later asked Mrs. Leonard directly what she knew about Detroit. She had heard of it, but knew nothing about it or even where it was in America. To check this further, Mrs. Allison questioned a dozen well-educated English people, and similarly all of them had heard of Detroit, but none of them knew it was the automobile manufacturing center of America (Allison, 1934:134). We also should keep in mind that this was 1929, when the automobile industry was relatively new and worldwide communication was not what it is today. And, of course, we should not forget that in any case Mrs. Leonard had no way of knowing normally where Mr. Thomas and his wife lived.

Another interesting feature of the John Thomas series of sittings is the occurrence again of cross-references to the same topics by different mediums who knew nothing about each other, as in the cross-correspondence cases and in the Macaulay and Edgar Vandy cases. Thomas devotes an entire chapter of his book to this topic (Thomas, 1937:10, 166–188). I will briefly mention one example here, which involved three mediums at seven sittings with three different sitters, over a two-and-a-half-year period. Although only one of these seven sittings was a proxy sitting, I want to describe this case as an example of how research involving both proxy and regular sittings might be combined to produce good evidence of cross-references or cross-correspondences.

On July 6, 1926, Mrs. Soule in Boston said: "She [i.e. the deceased Mrs. Thomas] holds up her hand. A ring drops off her finger. Test—test—lost gift—special gift—special occasion—to bind us together. She wishes she had another one now to bind us together" (Thomas, 1937:174). The facts were that Mrs. Thomas, at the time of her older son's engagement, took the stone from her own engagement ring and had it reset for her new daughter-in-law; but the ring was unfortunately lost before Mrs. Thomas's death.

Two months later, this time in London, another medium, Mr. Austin, said: "The other lady [again, meaning Mrs. Thomas] left a ring in possession of her sister, a lady of abundant dark hair pulled back from her forehead, pale skin, very slight build" (Thomas, 1937:174). The facts here were that, after Mrs. Thomas had had the stone removed from her engagement ring, she gave the setting to one of her sisters, her only sister who fit the description given. Neither Thomas's younger son, who was the sitter on this occasion, nor Thomas himself knew that she had done this.

Nearly five months later, again back in Boston, Mrs. Soule said: "Now I want to say a word about some of the tests I tried to send or give at other places—across the water there was a . . . little matter which seemed pretty good—to me— . . . RING . . . Yes, the engagement ring. It was what might be called a cross reference" (Thomas, 1937:174–175). This referred to the mention of the ring five months earlier, across the water, in London, by Mr. Austin.

A year later, in London, a third medium, Mrs. Garrett, said:

> Now she [Mrs. Thomas] speaks of a ring, a plain ring, but she also speaks of another ring which I think she had meant to go to a daughter or a daughter-in-law. It was a ring of sentimental interest but also had value, and is set with diamond. (Thomas, 1937:175)

Such a statement, taken alone, would not be terribly interesting since most women probably have rings, even diamond rings, that they would like to leave to a daughter or daughter-in-law; but in connection with the other references to a ring by other mediums, it becomes more interesting.

In two sittings several months later, Mrs. Garrett again referred to a ring, including the statement: "It has a precious setting, but it was an old fashioned setting, and that ring seems to have belonged to somebody through marriage now, the ring evidently having been altered" (Thomas, 1937:175).

Finally, two and a half years after her first statement about a ring, Mrs. Soule in Boston made her third reference to the ring, saying: "It seemed like a mysterious disappearance. It was taken by someone with an idea of doing something with it and it was never done but it was never put back in the same place" (apparently a reference to the ring's loss, although since no one knew how it had been lost, the detail here could not be verified). When John Thomas, the sitter on this occasion, asked, "Which ring?", the communicator replied: "One with the stone, diamond, which was removed for another reason. I think you know why that was done and I was happy having it done that way. It was a sentimental idea of mine and you were in that, too. It was our engage—wait a minute—all in the same spirit of our earliest love" (Thomas, 1937:175–176).

Later Proxy Research

Despite the extremely interesting and apparently successful proxy research carried out and reported in the 1920s and 1930s, this line of research, like most research with mediums, died away. Since that time, only a few studies have been reported, most of them not particularly successful.

West (1949) carried out a series of proxy sittings with 18 mediums, and in this series he introduced the method of providing sitters with not only their own reading but with several others, to see whether they could correctly, and blindly, pick their own. The results were not significant.

In 1953, five proxy sittings were held with the medium Arthur Ford. The five real sitters, not knowing which was their own, were asked to rate all of the statements on all five sittings, to see whether they would mark more statements as correct in their own sitting than in the others. Again, the results were not significant (MacRobert, 1954).

In 1958 Gertrude Schmeidler reported a study with the medium Mrs. Chapman in which there were four different proxy sitters and four different real sitters. Each proxy held a sitting with Mrs. Chapman for each real sitter, making a total of 16 sittings in all. As in the MacRobert study, the four sitters, not knowing which were their own readings, were asked to mark the accuracy of all the statements in all 16 sittings, four of which were their own and 12 of which were not. Schmeidler and one of her graduate students used the Pratt–Birge method (Pratt & Birge, 1948) to try to give probability values to the individual statements. Using Schmeidler's values, the results were significant, but using the graduate student's values, the results were only suggestive; and when a score was assigned simply on the basis of the number of items checked correct, without any probability weightings, the results were not significant (Schmeidler, 1958).

In 1966 Osis reported a complicated series of what he called "linkage" experiments. As in the Macaulay case 30 years earlier, Osis set up these proxy sittings so that there were one or more levels of intermediary persons between the real sitter and the proxy sitter, again in an attempt to make the hypothesis of telepathy more difficult. Although there were some interesting individual items, overall the results were not impressive (Osis, 1966).

Proxy research is not without its shortcomings. As Stevenson (1968) pointed out, "remov[ing] the sitter from the medium's presence . . . may diminish the motives of both the medium and [the deceased] communicator for communicating" (Stevenson, 1968:336). Stevenson himself, therefore, with his colleague Erlendur Haraldsson, introduced a variation on proxy research in which the sitter was present, but visually and acoustically isolated from the medium, and the experimenter sitting with the medium was blind as to the identity of the sitter (Haraldsson & Stevenson, 1974). Ten sitters participated. Each of them was given all ten readings, not knowing which was the one intended for them. They were asked to rank them as to how well they applied to themselves. Four of the ten sitters ranked their own reading #1[6] ($p < .01$), and two additional sitters ranked their reading #2.[7]

Unfortunately, more recent studies like that of Haraldsson and Stevenson

(Haraldsson & Stevenson, 1974) have not been as successful, producing results that were either not significant (Jensen & Cardeña, 2009, O'Keeffe & Wiseman, 2005, Schwartz, Geoffrion, Jain, Lewis, & Russek, 2003) or only marginally significant (Beischel & Schwartz, 2007).

A colleague and I recently conducted some research involving proxy sittings that was successful. In a first pilot study with four mediums and 12 sitters, I served as the proxy. Scoring on an item-by-item basis, the sitters blindly evaluated their own reading as well as three control readings. The results were not significant. In a second, larger study with nine mediums and 40 sitters, my colleague and I each served as proxy for 20 readings. The sitters blindly evaluated their own reading as well as five control readings, but in this study they were asked to do so globally, rating each reading on a scale of 1 to 10. The results were highly significant ($p < .0001$) (for details about this research, see E. W. Kelly and D. Arcangel, An investigation of mediums who claim to give information about deceased persons, unpublished).

Future Research

This last study suggests that we can again produce significant results with proxy sittings, research that might eventually put us in a better position to develop new ideas for evaluating the survival hypothesis. Numerous questions suggested by previous investigators should be followed up on, particularly those addressing the issue of identifying the most conducive conditions for obtaining veridical information supernormally, *whatever* the source. For example (and these suggestions are given in no particular order):

1) Should the proxy be *completely* blind, which of course improves the evidentiality? Or does it help for the proxy to have some minimal knowledge or contact with the deceased person's survivors, perhaps to "prime the pump"? William James, for example, believed that the latter might be the "best policy. For it often happens, if you give this trance personage a name or some small fact for the lack of which he is brought to a standstill, that he will then start off with a copious flow of additional talk, containing in itself an abundance of 'tests'" (James, 1890:652).

2) Stevenson (1968) had suggested that the presence of the real sitter might increase the motivation for a deceased person to communicate, and in Saltmarsh's study (1929) results were better when the sitter was present than during proxy sittings. Can we examine this question of motivation, yet keep the methodological advantages of proxy studies, by somehow combining proxy and non-proxy sittings, either in different sittings as Saltmarsh did, or simultaneously as Haraldsson and Stevenson (1974) did?

3) Gardner Murphy believed that "the most cogent type of survival evidence" would be that suggesting "post-mortem *interaction* of two or more communicators," and he offered his hypothetical example of several people, unknown to each other in life, who learn after death that they all had some shared interest, such as collecting old Wedgwood china, and then communicate this fact through a medium (Murphy, 1945b:208). Some cross-correspondences, of course, seemed to be of this type[8], but can we find a way to produce similar evidence without waiting for someone else (deceased or living) to take the initiative? Nea Walker had used a kind of "appointment" or "invitation" method. Because she had been sitting with Mrs. Leonard for many years, a group of her own deceased friends and relatives had emerged as regular communicators, comparable to Mrs. Leonard's regular trance personality, Feda. They served as a kind of master of ceremonies, bringing other deceased people to the sitting and, often, conveying their messages for them. Before a sitting at which she wanted a particular person to appear, Miss Walker would appeal to her Group, asking that they try to find a particular person and bring this person to a designated sitting. Can we modify this "invitation" method to encourage the kind of interaction that Murphy suggested?

4) Can we identify some factor or factors correlated with successful communications? For example, is the age or character of the deceased person a factor, as Drayton Thomas (Thomas, 1932–1933) suggested? Is the mode of death a factor, as it seems to be in cases of the reincarnation type (Stevenson, 1987/2001:165–166) and apparition cases (Stevenson, 1982:346–347), in which violent or sudden death figures prominently? Even if the real sitters are not present, does it make a difference whether they do or do not consciously know that a sitting is taking place? Does the proxy sitter make a difference? (In the study by Kelly and Arcangel, there was no significant difference between the two proxies.)

5) Some of the best mediums of the past have used a "token" object belonging to the deceased person as their preferred means of establishing contact with the intended deceased person (Gauld, 1982:132, Saltmarsh, 1929). Does the use of such objects improve the results?

6) Are the results better when the mediums go into trance (or some other altered state)? Most of the best mediums of the past were trance mediums, whereas few mediums today seem to be.

Clearly, there is much that we could do to advance mediumship research both methodologically and theoretically. A list of questions such as that above could be extended indefinitely, but this is a useless exercise unless we can identify mediums able and willing to work under proxy conditions. Many otherwise good mediums may in fact not be successful under such difficult conditions; but some, like Mrs. Leonard, may find that they can do well. If so, it is particularly important that they are also willing to contribute to a sustained effort. Stevenson (1968:335) cautioned against expecting too much from one sitting. As he put it:

> Would a modern psychologist or psychiatrist usually expect to elicit information of a highly intimate nature about the life of a patient on the very first interview? . . . parapsychologists often err by expecting significant data to emerge at a first or only sitting instead of arranging for a series of sittings . . . we should expect evidence to emerge from investigations in which one or more sympathetic investigators arrange for the same sitter or sitters to participate in a rather long series of regular sittings.

One of the strongest impressions that I take away from the reports of the proxy sittings in the 1920s and 1930s is how much of a team effort this research was—on the part of the mediums, who were willing to persist over a period of years in cooperating with investigators; on the part of the investigators, who were also the proxy sitters and who developed a close and congenial relationship with the mediums by sitting with them regularly for years; and on the part of bereaved sitters who, as Oliver Lodge (1935:11) noted, understood "the importance of an outlook wider than their own immediate sorrow and need." With such a collaborative effort, we may yet again produce important evidence from mediumship, perhaps even some that could ultimately move us beyond the survival/super-psi impasse.

Notes

[1] I am drawing the distinction here between parapsychology, or the experimental study of psi, and psychical research, which I take to be a broader approach, as conceived by its founders, to general questions about the relationship of mind and body.
[2] Stevenson (1970:53) mentioned that he had collected about 60 published reports of drop-in cases and was preparing a monograph on them. Unfortunately, by the time of his death he had not done this. I am working on the material that he left and plan to publish a report and analysis of the cases.
[3] It did not seem to me that the readings improved over time, as Nea Walker's knowledge about the Whites increased; and Miss Walker herself thought that the best evidential material came in her proxy sittings, not in the sittings Mrs. White had (anonymously) with Mrs. Leonard, at which Mrs. White could have provided feedback (Walker, 1927:154). Nevertheless, a thorough analysis of this question would be of interest.
[4] Today audio and even video tape recordings should routinely be made as part of the effort to eliminate "cold reading" as an explanation for correct details.
[5] This last sentence may not have been entirely correct. Bobbie's friend Jack knew about the pipes and might have had some concern that the water there had been a factor in Bobbie's illness.
[6] In one of these four cases, the subject, a young woman, did not recognize anything

in what turned out to be her readings, and so she herself did not pick the correct reading. Nevertheless, when Haraldsson, still unaware of the identity of each sitter's actual reading, took all the readings to her family for their judgment, older family members immediately identified the correct reading as having numerous names and other details correct for persons that the subject was too young to have known.

[7] One other subject ranked the correct reading #5, but the other three did not rank their reading at all. Most sitters ranked only two to five readings, saying that the others had nothing meaningful enough to allow a judgment. The authors were thus unable to do a full-rank analysis, as they had planned. Similarly, a sum-of-ranks analysis, using only the ranking given to the correct reading (Solfvin, Kelly, & Burdick, 1978), is also not now possible.

[8] See, for example, the "Ear of Dionysius" case (Balfour, 1918; for a summary, see Murphy, 1961:252–270).

Acknowledgements

I wish to thank Stephen Braude and Edward F. Kelly for comments that have helped me greatly in revising this paper.

References

Allison, L. W. (1934). Proxy sittings with Mrs. Leonard. *Proceedings of the Society for Psychical Research, 21,* 104–145.
Balfour, G. W. (1918). The Ear of Dionysius: Further scripts affording evidence of personal survival. *Proceedings of the Society for Psychical Research, 29,* 197–243.
Beischel, J., & Schwartz, G. E. (2007). Anomalous information reception by research mediums demonstrated using a novel triple-blind protocol. *Explore, 3,* 23–27.
Bem, D. J., & Honorton, C. (1994). Does psi exist? Replicable evidence for an anomalous process of information transfer. *Psychological Bulletin, 115,* 4–18.
Braude, S. E. (2003). *Immortal Remains: The Evidence for Life after Death.* Lanham, MD: Rowman & Littlefield.
Broad, C. D. (1962). *Lectures on Psychical Research.* New York: Humanities Press.
Burdick, D. S., & Kelly, E. F. (1977). Statistical methods in parapsychological research. In: B. B. Wolman (Ed.), *Handbook of Parapsychology* (pp. 81–130). New York: Van Nostrand Reinhold.
Cook [Kelly], E. W. (1987). The survival question: Impasse or crux? *Journal of the American Society for Psychical Research, 81,* 125–139.
Cook [Kelly], E. W. (1992). *Frederic W. H. Myers: Parapsychology and Its Potential Contribution to Psychology.* [Unpublished Ph.D. dissertation]. University of Edinburgh.
Gauld, A. (1961). The "super-ESP" hypothesis. *Proceedings of the Society for Psychical Research, 53,* 226–246.
Gauld, A. (1968). *The Founders of Psychical Research.* London: Routledge & Kegan Paul.
Gauld, A. (1971). A series of "drop-in" communicators. *Proceedings of the Society for Psychical Research, 55,* 273–340.
Gauld, A. (1982). *Mediumship and Survival: A Century of Investigations.* London: Heinemann.
Gay, K. (1957). The case of Edgar Vandy. *Journal of the Society for Psychical Research, 39,* 2–64.
Haraldsson, E., & Stevenson, I. (1974). An experiment with the Icelandic medium Hafsteinn Björnsson. *Journal of the American Society for Psychical Research, 68,* 192–202.
Haraldsson, E., & Stevenson, I. (1975a). A communicator of the "drop in" type in Iceland: The case of Gudni Magnusson. *Journal of the American Society for Psychical Research, 69,* 245–261.

Haraldsson, E., & Stevenson, I. (1975b). A communicator of the "drop in" type in Iceland: The case of Runolfur Runolfsson. *Journal of the American Society for Psychical Research, 69,* 33–59.

Hyslop, J. H. (1919). Chance coincidence and guessing. *Proceedings of the American Society for Psychical Research, 13,* 5–88.

James, W. (1890). A record of observations of certain phenomena of trance (Part III). *Proceedings of the Society for Psychical Research, 6,* 651–659.

Jensen, C. G., & Cardeña, E. (2009). A controlled long-distance test of a professional medium. *European Journal of Parapsychology, 24,* 53–67.

Johnson, A. (1908). On the automatic writing of Mrs. Holland. *Proceedings of the Society for Psychical Research, 21,* 166–391.

Kelly, E. F., Kelly, E. W., Crabtree, A., Gauld, A., Grosso, M., & Greyson, B. (2007). *Irreducible Mind: Toward a Psychology for the 21st Century.* Lanham, MD: Rowman & Littlefield.

Lodge, O. (1935). Foreword. In: N. Walker, *Through a Stranger's Hands* (pp. 9–11). London: Hutchinson.

Mackenzie, A. (1971). An "Edgar Vandy" proxy sitting. *Journal of the Society for Psychical Research, 46,* 66–173.

MacRobert, A. F. (1954). Proxy sittings: A report of the study group series with Arthur Ford. *Journal of the American Society for Psychical Research, 48,* 71–73.

Murphy, G. (1945a). Difficulties confronting the survival hypothesis. *Journal of the American Society for Psychical Research, 39,* 67–94.

Murphy, G. (1945b). Field theory and survival. *Journal of the American Society for Psychical Research, 39,* 181–209.

Murphy, G. (1961). *Challenge of Psychical Research: A Primer of Parapsychology.* New York: Harper.

Myers, F. W. H. (1884). On a telepathic explanation of some so-called Spiritualistic phenomena. *Proceedings of the Society for Psychical Research, 2,* 217–237.

Myers, F. W. H. (1903). *Human Personality and Its Survival of Bodily Death.* London: Longmans Green.

O'Keeffe, C., & Wiseman, R. (2005). Testing alleged mediumship: Methods and results. *British Journal of Psychology, 96,* 165–179.

Osis, K. (1966). Linkage experiments with mediums. *Journal of the American Society for Psychical Research, 60,* 91–124.

Pratt, J. G. (1936). *Towards a Method of Evaluating Mediumistic Material (Bulletin 23).* Boston: Boston Society for Psychic Research.

Pratt, J. G., & Birge, W. R. (1948). Appraising verbal test material in parapsychology. *Journal of Parapsychology, 12,* 236–256.

Saltmarsh, H. F. (1929). Report on the investigation of some sittings with Mrs. Warren Elliott. *Proceedings of the Society for Psychical Research, 39,* 47–184.

Saltmarsh, H. F. (1938). *Evidence of Personal Survival from Cross-Correspondences.* London: G. Bell & Sons.

Saltmarsh, H. F., & Soal, S. G. (1930). A method of estimating the supernormal content of mediumistic communications. *Proceedings of the Society for Psychical Research, 39,* 266–271.

Schmeidler, G. (1958). Analysis and evaluation of proxy sessions with Mrs. Caroline Chapman. *Journal of Parapsychology, 22,* 137–155.

Schouten, S. A. (1993). Applied parapsychology: Studies of psychics and healers. *Journal of Scientific Exploration, 7,* 375–401.

Schouten, S. A. (1994). An overview of quantitatively evaluated studies with mediums and psychics. *Journal of the American Society for Psychical Research, 88,* 221–254.

Schwartz, G. E., Geoffrion, S., Jain, S., Lewis, S., & Russek, L. G. (2003). Evidence of anomalous information retrieval between two mediums: Replication in a double-blind design. *Journal of the Society for Psychical Research, 67,* 115–130.

Solfvin, G. F., Kelly, E. F., & Burdick, D. S. (1978). Some new methods of analysis for preferential-ranking data. *Journal of the American Society for Psychical Research, 72,* 93–110.

Stevenson, I. (1968). The analysis of a mediumistic session by a new method. *Journal of the American Society for Psychical Research, 62,* 334–355.

Stevenson, I. (1970). A communicator unknown to medium and sitters: The case of Robert Passanah. *Journal of the American Society for Psychical Research, 64,* 53–65.

Stevenson, I. (1982). The contribution of apparitions to the evidence for survival. *Journal of the American Society for Psychical Research, 76,* 341–358.

Stevenson, I. (1983). Cryptomnesia and parapsychology. *Journal of the Society for Psychical Research, 52,* 1–30.

Stevenson, I. (2001). *Children Who Remember Previous Lives: A Question of Reincarnation* (rev. ed.). Jefferson, NC: McFarland. (Original work published 1987)

Stevenson, I., & Beloff, J. (1980). An analysis of some suspect drop-in communicators. *Journal of the Society for Psychical Research, 50,* 427–447.

Sudduth, M. (2009). Super-psi and the survivalist interpretation of mediumship. *Journal of Scientific Exploration, 23,* 167–193.

Thomas, C. D. (1932–1933). A consideration of a series of proxy sittings. *Proceedings of the Society for Psychical Research, 41,* 139–185.

Thomas, C. D. (1935). A proxy case extending over eleven sittings with Mrs. Osborne Leonard. *Proceedings of the Society for Psychical Research, 43,* 439–519.

Thomas, C. D. (1938–1939). A proxy experiment of some significant success. *Proceedings of the Society for Psychical Research, 45,* 257–306.

Thomas, C. D. (1939). A new type of proxy case. *Journal of the Society for Psychical Research 31,* 103–106, 120–123.

Thomas, J. F. (1937). *Beyond Normal Cognition: An Evaluative and Methodological Study of the Mental Content of Certain Trance Phenomena.* Boston: Boston Society for Psychic Research.

Verrall, Mrs. A. W. (1906). On a series of automatic writings. *Proceedings of the Society for Psychical Research, 20,* 1–432.

Walker, N. (1927). *The Bridge: A Case for Survival.* London: Cassell.

Walker, N. (1929). The Tony Burman case. *Proceedings of the Society for Psychical Research, 39,* 1–44.

Walker, N. (1935). *Through a Stranger's Hands.* London: Hutchinson.

West, D. J. (1949). Some proxy sittings: A preliminary attempt at objective assessment. *Journal of the Society for Psychical Research, 35,* 96–101.

HISTORICAL PERSPECTIVE

Parapsychology in France after May 1968:
A History of GERP

RENAUD EVRARD

Department of Psychology, University of Rouen, France
evrardrenaud@gmail.com

Abstract—The Group for the Study and Research of Parapsychology (GERP) (*Groupe d'Etudes et de Recherche en Parapsychologie*) was originally composed of psychology undergraduates who, in the context of the period after May 1968, tried to bring into the curriculum a course of parapsychology at the University of Paris X Nanterre. Their failure transformed the group into an association of young researchers with much solidarity, criticizing foreign models and developing a theory-oriented parapsychological cross-disciplinary research in the 1970s and 1980s. The works of François Favre and Pierre Janin, two influential members of GERP, are reviewed as examples. A short presentation is offered of some of their theories. GERP although dormant has not been dissolved, thus this history reveals a recent but underestimated period of French parapsychology and the lineaments of an original research to be deepened.

Keywords: history of French parapsychology—*Groupe d'Etudes et de Recherches en Parapsychologie* (GERP)—May 1968—theoretical parapsychology—sociology of science—fantastic realism

Parapsychology at the University

In the history of France during the twentieth century, May 1968 stands as the most important social movement. Launched by a revolt of Parisian students, the crisis was at once cultural, social, and political. We can also consider this moment as a caesura in the history of French parapsychology. During the late 1960s, the old and once-esteemed psychical research tradition of the Institut Métapsychique International (IMI), founded and "recognized for public interest" by the French state in 1919, was in the midst of a long period of decline and lacked the necessary funds to revive its activities in response to the interest of a younger generation of scholars. For its part, the literary movement incorporating occult studies named "fantastic realism", which had been launched by the publication of *Le Matin des Magiciens* (Pauwels & Bergier, 1960) and sustained in the journal *Planète*, generally remained ambivalent with regard to the *événements* of May and provided little guidance to students interested in the study of the paranormal (Gutierez, 1998). The fantastic realism movement was highly influential in the way it presented occult but also parascientific subjects to a popular audience. It encouraged many cultural breakthroughs through a change in spiritual and intellectual awareness, preparing in some ways the ideas of the student movement and especially openness to the paranormal. But the disappearance of *Planète* in 1968 effectively signaled the end of the movement.

Given the lack of an established institutional base for the study of parapsychology, a new generation of researchers would take it upon themselves to create the Group for the Study and Research of Parapsychology (*Groupe d'Etudes et de Recherches en Parapsychologie,* or GERP). The anarchic nature of the organization was well-adapted to the political ideas of the student movement, which seemed to have a naturally given entree to the paranormal and the parapsychological in accordance with the sociological pattern of the "Trickster", put into evidence by Hansen (2001): A chaotic structure in a revolutionary context seems to have more affinity for becoming interested in scientific margins.

The original impetus for the organization had come from psychology undergraduates who, intrigued by the work of the American researcher J. B. Rhine, were quietly encouraged in their interests by Rémy Chauvin, professor of animal behavior at the University of Paris V. The students found more public support for their interest in parapsychology via a national television program featuring the German researcher Hans Bender, from whom they learned more about the current laboratory efforts in parapsychology taking place in Germany. Inspired by the German example, about twenty students at the University of Paris X (Nanterre) created an informal association, called the Group for the Study and Research of Metapsychics (GERM)[1].

Enthusiasm within the two groups was high and goals ambitious. The students' first step was to attempt to bring parapsychology into the curriculum as a subdiscipline of psychology by petitioning their professors to create an officially accredited course. These activists, some still in their teens, were ignorant of the administrative challenges they faced. At the end of more than two years of effort, many would leave the organization demoralized and resentful.

Resistance among the faculty was high. Some professors, influenced by the skeptical literature on parapsychology, considered the field a pseudoscience. Others thought that its introduction into the curriculum would compromise the entire department. A few teachers were, however, curious enough to consent to organizing a series of meetings to consider the question further.

On January 25, 1971, the physicist Henri Marcotte gathered 400 students in one such meeting, but his demonstration, which involved a collective telepathy experiment, rapidly turned in a farce (personal communication from Favre, June 2007). Since they came to discover scientific parapsychology, the attendees were not ready to proceed directly to personal practice. The lack of distance between Marcotte and the phenomenon of telepathy had discredited his discourse on the objectivity and scientificity of parapsychology. The second opportunity would come in May, with the intervention of Hans Bender, but the audience in this instance was much smaller. Only ten professors out of 54 attended his presentation. Through his professionalism and his humanity, Bender embodied the model researcher in the eyes of the students. Regardless, following the two events, university administrators judged in a meeting in October 1971 that there would be no course in parapsychology. They did concede, however, to provide on a temporary basis a small room for the realization of experiments and to sponsor another public talk on parapsychology, this time by Rémy Chauvin.

The debate among the faculty remained contentious, as reported by the disappointed students in *Revue Métapsychique* (GERP, 1971–1972). According to them, some teachers claimed they had collaborated with Rhine and could now pronounce parapsychology dead because Rhine himself had refuted in print all of his earlier conclusions! Supported by professors Rhine, Musso, Bender, and Chauvin, the proposal drawn up by the students was sent personally to every professor and was generally ignored. Professors remained silent when students challenged them to personally review the experiments that they had conducted at the university on a regular basis since 1970. In addition to replications of Rhine's ESP research, the students had also succeeded in reproducing psi experiments with mice, which Chauvin and Mayer had published under the pseudonyms Duval and Montredon in the *Journal of Parapsychology* (Duval & Montredon, 1968a, 1968b), and for which they received the McDougall Award in 1970. Annoyed at the faculty's response, the students demanded a clear answer.

A small contingent of students traveled to Bender's lab at the Institute of Border Areas of Psychology and Mental Hygiene (IGPP) in Freiburg, Germany, and discovered that research in parapsychology was possible within the institutions of the academy. Inspired by the German example, GERM abandoned the antiquated term "*la métapsychique*," and adopted the more reputable term "parapsychology." As GERP, the organization was recognized as a legal association in July 1971. François Favre, who joined the group that year and soon became its leader, insisted that the vocation of "mental hygiene" (by which he meant something equivalent to the current clinical psychology of exceptional experiences) be directly registered in the statutes of the association as in the German model. According to this model, parapsychology was not reducible to a laboratory-based approach of quantifiable phenomena. The starting point remained the paranormal experiences reported by the population and their analysis by the humanities. The parapsychologists had a sociological and clinical role to play in disseminating scientific information about these experiences, which could help those who suffered from them. This more human-centered approach, which highlighted the social role of parapsychology, corresponded more with the attitude of GERP researchers.

Through press accounts of these various challenges to the state and university, the association earned notoriety among French students. The question that remained was whether or not its research agenda could take place within the university. According to the report of GERP's members (GERP, 1971–1972:98–99), a petition signed by 183 psychology students (90% of those solicited) indicated that the student population was in favor of creating a one-credit course in parapsychology. However, the president of the administrative council "forgot" to bring with him to the meeting of April 26, 1972, the petition that the students had presented to him. The administration similarly "forgot" to mention this petition and its terms in the official report of the meeting. The efforts of the students were thwarted at each step by an administration that simply postponed the question. It was not until the board meeting of November 1972, after more than two years of student effort that the administration finally responded to their demands. Parapsychology, they concluded, interested no one and did not enter into the framework of the discipline of psychology. The rejection by the Nanterre administration led GERP members to take their demands elsewhere and, indeed, many of these members would be the initiators of elective courses in parapsychology at other French universities and in high schools.[2] In spite of these small gains, GERP suffered as a result of both its lack of strategic subtlety and the general decline of student idealism after 1968. In fact, only three members of the organization would survive this initial period. In their hands, GERP would move away from direct engagement with the university and focus on developing a theoretical framework for French parapsychology.

The Researchers

In this new phase of activity, GERP abandoned Rhine's experimental paradigm in favor of developing cross-disciplinary approaches to psi phenomena. Around François Favre, who had psychiatric training and chose to dedicate his life to this domain, generations of young researchers would gather, investigating various aspects of parapsychology. These included mental hygiene and psychoanalysis (Nicole Gibrat, Francis Danest, Pascal Le Maléfan, Christian Moreau, Pascale Catala, Johann Mathieu), physics (Pierre Janin, Christian Cabayé, and polytechnicians Michel Duneau, Hervé Gresse, and Georges Nicoulaud), philosophy of sciences (Pascal Michel, Marc Beigbeder), ethnology (Philippe Léna), biology (Guy Béney, Pascal Lemaire), and even ufology (Pierre Viéroudy, Jacques Vallée, Bertrand Méheust). In the work of these scholars, parapsychology was reconceived as a field in which all scientific disciplines were deemed relevant and in which no single discipline would be sufficient. The approach was successful thanks to the great solidarity among group members, colleagues, and friends, and to the volunteer activities of individuals such as Gisèle Titeux and Janine Rousselier.

Two researchers were particularly active: Favre excelled in the cultural fields, and Janin in the natural sciences. Favre began by creating a historical account that reconstructed the genealogy of western parapsychology (Favre, 1975, 1976). After concluding this chronological account, he created a geographical analysis by compiling data from 450 surveys of world beliefs (Favre, 1988). He also became an expert on ectoplasm materialization studies (Favre, 1973), which he described in a complementarist perspective as a "materialized dream".[3] He then made a link between UFO and psi apparitions (Favre, 1996), at the origin of a close debate of parapsychologists and ufologists (such as Vallée, Viéroudy, and Méheust). He was also a specialist on onirism, i.e. issues related to dream functioning (Favre, 1998). As a prolific theorist, Favre elaborated a model, regrettably overlooked in other fields, that attempted to resolve by a complementarist logic the paradoxes of psi with regard to time (Favre, 1982), and also psychosomatic phenomena (Favre, 1995). He challenged all those who claimed the title of "parapsychologist" to engage in debate, but the complexity of his theoretical model and the aggressive nature of his exchanges discouraged many of his potential interlocutors.[4] Favre would, in many ways, remain a "sixty-eighter" in the insistent nature of his idealism.

Pierre Janin was a brilliant engineer and also a holder of a degree in philosophy when he came to Professor Rémy Chauvin's animal behavior laboratory in 1969 to study parapsychology. He performed PK experiments with larvae which he subsequently presented to Rhine in the summer of 1969.[5] He dedicated eight hours per day to parapsychology, treating it as his profession even though he was not paid. This commitment allowed him to become an expert in the parapsychological literature and to acquire an international reputation. Janin's spirit of dedication and sacrifice resembled those of the activists of 1968, with parapsychology serving as another way of changing the world, in this case at the metaphysical level. This motivation is clear in Janin's major theoretical article entitled "New Perspectives on the Relations between the Psyche and the Cosmos" (Janin, 1973). Sent to Helmut Schmidt, this text would inspire the first experiments on retro-PK. This article contained the germ of Janin's entire research agenda—to approach randomness as a psychophysical phenomena, i.e. that it is not absurd but meaningful physical coincidences under the mental influence of a living individual or group (the so-called "neo-animistic theory" with regard to the place of mind in nature); and to test traditional mantics (based on interpretation of chaotic forms) in comparison with modern diagnostic methods.

The first part of this program was pursued in retro-PK experiments sponsored by the Parapsychology Foundation in New York City (Janin, 1976). These were less sophisticated and convincing than those of Schmidt, but were important in developing the *tychoscope*, a randomly moving robot designed to test if PK was like a subjective relationship with an object, as if both the subject and the object could unite in a dyad (Janin, 1977). The first copy was conceived in June, 1975, and 24 copies were available in 1981. For this project, Janin again benefited from a grant from the Parapsychology Foundation, but also from other important financial resources, most notably the famous French industrialist Ambroise Roux. These experiments on "the nature of randomness" didn't confirm Janin's neo-animistic conceptions (1975), but his device would play a crucial part in René Péoc'h's experiments (1986) in which the random movements of the tychoscope were "imprinted" by baby chicks. Janin would also conduct, with less support from his colleagues, experiments designed to extend the ideas of C. G. Jung by comparing the traditional mantic of astrology with psychiatric diagnosis (Janin, 1976). Janin was the only member of GERP[6] who was this active in experimentation, which he would justify as a necessary complement to the theoretical approaches emphasized by his fellows (Janin, 1979).

The general conviction of GERP, however, was that parapsychologists had already produced too much data, and that it was now a question of analyzing, synthesizing, and explaining them in a critical way. Many felt that it was useless to repeat the same proof-oriented psi experiments over and over again. If no scientific consensus had emerged after decades of paranormal research, thousands of supplementary experiments with Zener cards were unlikely to contribute to one in the present. As for the experiments on the conditions of the occurrence of psi, they were attributed to experimenter effects and only confirmed the theoretical presuppositions or the inner worldview of the experimenter.

Some Theoretical Postulates

In spite of several attempts, the members of GERP failed to reach a theoretical consensus on what psi was (Michel, Favre, Janin, Beney, & Hemmerlin, 1986). Nevertheless, following fruitful debates on all the facets of psi phenomena, some general propositions were offered. These propositions were diffused in journals to the wide or restricted public,[7] by numerous courses and lectures in Paris, by several conferences,[8] and by the Federation of Parapsychological and Psychotronic Research Groups (Fédération des Organismes de Recherche en Parapsychologie et Psychotronique, FOREPP). The latter group was founded in 1975 and intellectually dominated by GERP even though its aim was to encourage the activities of provincial groups. FOREPP tried in particular to bring practitioners of the paranormal (psychics, healers, dowsers, etc.) into a dialogue with scientists. The conflict between research and commercial activity, however, revealed itself to be insoluble. In spite of these efforts to publicize their ideas, GERP's theoretical constructs remained largely unknown to parapsychologists internationally. The refusal of the Rhinian experimental paradigm was upheld by the majority of the group's members in favor of cross-disciplinary approaches. The group's basic theoretical precepts included:

The experimenter effect is predominant: In any psi experiment, one cannot really distinguish the influences of the subject from those of the experimenter.

Psi phenomena are homogeneous: All psi phenomena can be encompassed by a single explanation.

ESP is auto-premonition (Favre, 1982): One can reduce all paranormal acquisition of information to the premonition of the feedback which will allow verification of this acquisition.

Every instance of PK is a retro-PK (Janin, 1973): Because quantum physics had shown that there are undetermined processes at the source of all physical events, PK could influence these processes without breaking thermodynamics laws, and then PK in the present can be reduced in theory to a PK in the past.

Synchronicity's model: Psi events cannot be described as causal transmissions of energy, but rather as "significant coincidences". For many years, several members of GERP followed guidelines drawn by Jung and Pauli, before trying to exceed them.

Psi implies a psychophysical complementarity: The psyche is irreducible to physics, and vice versa. Interpretations of psi phenomena only by physics or only by psychology are excluded from consideration. Numerous researchers lie nevertheless in this imbalance, becoming the propagators of physicalist myths of psi (Duneau, 1979, 1980). The complementarist approach of psi is supported by Stéphane Lupasco's non-Aristotelian logic of the contradictory, which goes beyond mind–matter dualism and does not deny parapsychological implications (cf. the last pages of Lupasco, 1979; and the thesis of his pupil and member of GERP, Beigbeder, 1977).

The challenges of proving psi in animals: Animal (or vegetal) psi cannot be distinguished from the psi phenomenon produced by human experimenters. This theoretical principle would mark the opposition of GERP members against the empirical deductions of Rémy Chauvin and his pupil René Péoc'h. The French research tradition on animal or vegetal psi was criticized on this basis: Because we can never exclude a human PK following the conscious or unconscious motives of the experimenter, it is not fair to attach biological and materialist preconceptions on these data, suggesting that there are psi sources other than humans.

The anti-scientific aspects of parapsychology: Parapsychology is explicitly scientific in its methods, but psi itself is impervious to cumulative scientific investigation. Psi produces a subversion of scientific investigation by its elusiveness, i.e. its tendency not to be exactly reproducible in an identical manner. This can be understood as an anti-scientific property. One finds here similarities with the occultist Robert Amadou, who had already stirred up ill feeling in the French parapsychological community in the 1950s. Amadou (1954) had harshly criticized the old parapsychology that didn't use the rigorous methods of Rhine but finally explained paranormal phenomena by occult or mystical theories.

The mythogenic part of psi: Psi is a major generator of myths, which are latent but radical subversions of daily materialist references. It thus obliges one to stand back, like an anthropologist, with regard to conceptions of what is "paranormal". Like the diabolic, spiritualist, and extraterrestrial hypotheses in human cultures, the hypothesis of psi-gifted subjects (like supermen or mutants) extends the function of myth within a culture influenced by science.

Psi and finality in evolution: Psi suggests reintroducing the hypothesis of neo-Lamarckian process in biological evolution. One example presented by Favre (2004) involved statistics of the postwar population in which the net increase of male births was significant. Favre speculated that this demographic pattern was produced in response to the need to replace the male population lost to the war. This fact, known by all the sociologists, remained unexplained by biology and could, he argued, imply a psi process.

The ambivalence of sheep–goat in the same subject (Janin, 1973): A classical distinction is made in parapsychology between "sheep"—those who were confident about the reality of psi—and "goats"—for those who doubted its existence or its pertinence in the context of a test. But, for GERP's members, there are not on one side sheep and on the other side goats: Every individual contains in himself his own sheep and his own goat. In this subjective contradiction, only one of the terms is consciously expressed (the psi wish), the other remains unexpressed but nevertheless effective, as "psi missing" shows.

Historical aspect of psi: There are no psi effects but only psi events ("psiphanies"), at the crossroads of history and science. Therefore, the historical approach has great importance in parapsychology, serving as it does as a means of struggling against the naive progressivism of parapsychologists who believe themselves to be looking forward while ignoring the controversies of the past.

Common and rare psi: According to François Favre, some common psychic processes work as "common" psi: memory, rational forecast, intentionality, etc. *"Make a remote movement of a table or a voluntarily movement of your own hand, it is exactly the same scientific issue,"* said Gasparin (1854/1888:125–126). It is thus necessary to consider the so-called paranormal phenomena as extremities of a continuum of process at work everywhere. "Paranormal" events are just *rare* events which reveal, on a large scale, our ignorance of the mechanism of some fundamental but *frequent* phenomena (mind–body relationship, intentionality, the nature of time and space, etc.). In this sense, Favre (2002) reinterpreted Benjamin Libet's experiments on the cerebral effect of decisionmaking as the reproducible proof of common retro-PK.

These propositions while innovative for this period also have some recognizable sources. There is notably C. G. Jung's work, then marginalized by the French psychoanalysts, but rehabilitated thanks to the seminal work of Henri Ellenberger (1970). More important still is the scientific essayist René Sudre (lived 1880–1968) (Evrard, 2009), whose work sought to reconcile the subjective and objective aspects of the metapsychics. The extension of psi investigation in a general study of nature was suggested by his book *The New Enigma of the Universe* (Sudre, 1943), in which Sudre brilliantly discusses the biggest scientific questions of the twentieth century. His encyclopedic discussion brings the reader to the following conclusion: that in every discipline enigmas persist that constitute ruptures in the fabric of knowledge and through which metapsychical questions come rushing in. Sudre argued that the goal of the psychical researcher must be to look at these questions that psychical research shares in common with the other sciences and then to widen these points of contact to the benefit of scientific understanding in general. This is very much the direction that the cross-disciplinary and anti-establishment vocations of the GERP members went.

Epilogue

In spite of what our historical account might suggest, GERP still remains in existence. While family and professional concerns have led to a dispersal of its members, most continue to pursue their isolated researches. Only a hard core of eternal students persisted. Some made successful careers as scholars at the academic level. Michel Duneau, for instance, worked as a research supervisor in physical theory at the National Centre for Scientific Research. While their ideas were not dissimilar from some of their contemporaries elsewhere in Europe, GERP had little influence abroad. The aim of this group evolved dramatically from its early days. Initially inspired by foreign examples in its efforts to impose parapsychology on the University, GERP would adopt a more strictly theoretical orientation in its efforts to bring parapsychology into a productive relationship with French traditions of inquiry. The boldness of GERP came in trying to reconceptualize psi in terms offered by other disciplines and to minimize the importance of the superficial conceptions of ESP and PK emphasized by the American approach. This boldness was a natural extension of the general cultural and intellectual ferment out of which French parapsychology would draw new life.

Acknowledgements

I wish to thank Matthew Brady Brower, Pascale Catala, François Favre, Pierre Janin, Pascal Michel, Guy Béney, Jean Roche, Pascal Le Maléfan, Georges Nicoulaud, and the *JSE* reviewers.

Notes

[1] Information about the first years of GERM/GERP was collected in an anonymous article (GERP, 1971–1972) and through numerous discussions with François Favre.
[2] In Paris VII (1973–1976), Tours (1976), HEC (1978–1981, 1983–1984), Supélec (1980–1982), and Angers (1989–1990).
[3] Favre also worked on two anthologies, on apparitions (Favre, 1978a) and on ectoplasms (Favre, 1978b), which had large public success, but for which the publisher did not pay him one coin.
[4] We can consult his articles and discussions at http://www.sciencesphilo.fr/
[5] These experiments were published under a pen name (Metta, 1972), following the advice of Chauvin.
[6] After failures in his experimental results, Janin resigned and chose a therapist's career to be able to at least "help locally".
[7] The journal *Parapsychologie* was sold in kiosks, and 26 isssues were published in a irregular way from 1971 until 1989. The *Bulletin psitt* addressed especially "psiphiles" in 47 issues that appeared from February 1982 until June 1986.
[8] Proceedings of these conferences were published in book collections (Collectif, 1976, Favre, 1992) of high standard, reporting discussions between lecturers and the public.

References

Amadou, R. (1954). *La parapsychologie. Essai historique et critique.* Paris: Denoël.
Beigbeder, M. (1977). *La clarté des abysses.* Paris: Morel.
Collectif (1976). *La parapsychologie devant la science.* Paris: Berg-Bélibaste.
Duneau, M. (1979). Aspects historiques des rapports entre parapsychologie et sciences exactes. *Parapsychologie, 7,* 11–18.
Duneau, M. (1980). Le mythe quantique en parapsychologie. *Parapsychologie, 11,* 2–7.
Duval, P. [Chauvin, R.], & Montredon, E. [Mayer, J.] (1968a). A PK Experiment with Mice. *Journal of Parapsychology, 32,* 153–166.
Duval, P. [Chauvin, R.], & Montredon, E. [Mayer, J.] (1968b). Further psi Experiments with Mice. *Journal of Parapsychology, 32,* 53–56.
Ellenberger, H. (1970). *The Discovery of the Unconscious. The History and Evolution of Dynamic Psychiatry.* New York: Basic Books.
Evrard, R. (2009). René Sudre (1880–1968): The Metapsychist's Quill. *Journal of the Society for Psychical Research, 73*(4), 207–222.
Favre, F. (1973). Psychorragies. In: Bender H, Chauvin R, Janin P, Larcher H. (Eds.), *60 années de parapsychologie* (pp. 271-292). Paris: Editions Kimé, 1992.
Favre, F. (1975). Les Enfants du Lac de Constance – 1ère partie. *Parapsychologie, 1,* 14–27.
Favre, F. (1976). Les Enfants du Lac de Constance – 2ème partie. *Parapsychologie, 2,* 11-32.
Favre, F. (1978a). *Les Apparitions mystérieuses.* Paris: Tchou.
Favre, F. (1978b). *Que savons-nous sur les fantômes?* Paris: Tchou.
Favre, F. (1982). Le modèle de l'auto-prémonition. *Parapsychologie, 14,* 9–29.
Favre, F. (1988). Le paranormal, la science et le monde. http://www.sciencesphilo.fr/
Favre, F. (Ed.) (1992). *60 années de parapsychologie.* Paris: Kimé.
Favre, F. (1995). Animisme et espace-temps. In: *Autour du rêve, Forum transdisciplinaire N°1.* Paris: Editions L'Atelier du rêve.
Favre, F. (1996). Ovnis et psi. http://www.sciencesphilo.fr/
Favre, F. (1998). Le rêve: Histoire, interprétation et psychothérapie. In: *Dossiers d'informations médicales et paramédicales* (Vol. 3). Paris: Livre de Paris, 1999.
Favre, F. (2002). Sur la parapsychologie et l'expérience de B. Libet. Correspondance avec Philippe Garnier. http://www.sciencesphilo.fr/
Favre, F. (2004). Psi et intentionnalité. Conférence donnée à l'IMI le 17 décembre 2004. http://www.sciencesphilo.fr/
Gasparin, A. de (1854/1888). *Des Tables tournantes, du surnaturel et en général des esprits, 2ème édition.* Paris: Calmann-Lévy.
GERP (1971–1972). La parapsychologie à l'université de Paris X Nanterre. *Revue Métapsychique, 17,* 95–100.
Gutierez, G. (1998). *Le discours du réalisme fantastique: La revue Planète.* Undergraduate work of Lettres Modernes Spécialisées, University of Paris IV.
Hansen, G. (2001). *The Trickster and the Paranormal.* Philadelphia: Xlibris.
Janin, P. (1973). Nouvelles relations entre la psyché et le cosmos. *Revue Métapsychique–Parapsychologie, 18,* 63–89.
Janin, P. (1975). Psychocinèse dans le passé? Une expérience exploratoire. *Revue métapsychique, 21–22,* 71–96.
Janin, P. (1976). Vers l'expérience répétable en parapsychologie: Le projet 'Gnomon'. In: Collectif, *La parapsychologie devant la science.* Paris: Berg-Bélibaste, pp. 94–121.
Janin, P. (1977). Psychisme et hasard. *Parapsychologie, 4,* 5–20.
Janin, P. (1979). A quoi sert aujourd'hui l'expérimentation en parapsychologie. *Parapsychologie, 8,* 3–18.
Lupasco, S. (1979). *L'univers psychique.* Paris: Denoël/Gauthier.
Metta, L. [Janin, P.] (1972). Psychokinesis on Lepidoptera larvae. *Journal of Parapsychology, 36,* 213–221.

Michel, P., Favre, F., Janin, P., Beney, G., & Hemmerlin, E. (1986). Propositions pour un consensus sur le psi. *Parapsychologie, 20,* 18–37.

Pauwels, L., & Bergier, J. (1960). *Le matin des magiciens: Introduction au réalisme fantastique.* Paris: Gallimard.

Péoc'h, R. (1986). *Mise en évidence d'un effet psychophysique chez l'homme et le poussin sur le tychoscope.* [Doctoral dissertation in Medicine]. University of Nantes.

Sudre, R. (1943). *Les nouvelles énigmes de l'univers.* Paris: Payot.

Most of the articles above are available at http://gerp.free.fr and at http://www.sciencesphilo.fr/

LETTERS TO THE EDITOR

Response to "How To Improve the Study and Documentation of Cases of the Reincarnation Type? A Reappraisal of the Case of Kemal Atasoy"

The Essay by Vitor Moura Visoni in *JSE, 24*(1), Spring 2010, pp. 101–108, makes a number of criticisms of our Research Article "Children Who Claim To Remember Previous Lives: Cases with Written Records Made before the Previous Personality Was Identified," *JSE, 19*(1), Spring 2005, pp. 91–101, which we will address by section:

The Participation of the Interpreter. We disagree that "such investigations have already suffered enough from the accusation of fraud on the part of the interpreters." One interpreter 40 years ago was accused of fraud in unrelated work, but his interviews were subsequently validated by other interpreters. When the child in the current case was interviewed, the interpreter's motives were irrelevant. Since the child was describing an obscure person from 50 years before, whose existence JK [the author] was only able to confirm after great effort, the interpreter could not reasonably be accused of falsifying the interview. At the time of the interview, he did not possess any information about the previous personality that he could have put in the mouth of the child.

The Interview with the Child. The author recommends recording all interviews. We have recorded interviews on occasion, but we agree with the concerns Dr. Haraldsson mentioned. Though having interviews recorded and transcribed might seem ideal, the process of getting them can be impractical, or worse can impact on the quality of the information being obtained.

The author also objects to the presence of the mother during the interview. We think anyone who has had experience with children would recognize that the chances of getting a six-year-old child to share information with strangers who do not speak the same language without a parent or close attachment figure present are extremely remote. In this case, the boy's mother could not have fed him any information since she knew nothing about the person being described.

Regarding the number of interviews, the boy and his family were interviewed multiple times, though we acknowledge that our paper could have been clearer on that point. The most important interview by far, however, was the first one, which was conducted before anyone tried to verify the child's statements. Dr. Haraldsson is right that multiple interviews can help ensure that there is consistency about what the child was alleged to have said before the case was solved. In this case, there is no question about what the child said before the case was solved, because JK solved it after interviewing the child.

We agree that finding as many witnesses to the child's statements as possible is helpful and often essential. In this case, the child had not made statements to anyone but his immediate family, as we stated in the article. More significantly, multiple witnesses are often needed to confirm that the child had the knowledge about the previous personality that his parents claim. In this case, that issue is not in doubt since the case was unsolved at the time of the initial interview.

The Interview with Mr. Toran Togar. The author argues that JK should not have been the one to conduct the confirmatory interview. When JK was searching for people who could tell about the history of the home in question, it would have been impractical to locate an informant and then say the interview would have to wait until another researcher could be flown in from another country, a researcher who would not know what to ask in order to confirm the boy's statements. That issue aside, the author is correct that a recording would serve as stronger confirmation of the interview than notes alone, but again there are practical issues as to why we do not routinely record interviews.

Tests of Recognition. The author faults the lack of recognition tests in this case. As we noted in the paper, the boy was beginning to forget details of the purported life by the time his statements had been verified. That and the changes that would have occurred in the city during the 50 years following the previous personality's death meant that the slim chance that he would be able to recognize places was outweighed by the factors, such as the expense to the investigation and the time required of the boy and his parents to travel 850 km each way, that made such a trip impractical. We do agree that if recognition tests are to be performed, they need to be tightly controlled to be of significant value.

Psychological Tests. We are unaware of any psychological disorder that could lead a child to know numerous details about a man who lived 850 km away and died 50 years before.

Description of the Case . . . . Though we do not have a verbatim transcript of the interviews that were conducted, we do provide a list in our paper of all the statements the boy made before any attempt was made to verify them.

Is It Still a Strong Case? On this point, we are in full agreement with the author's positive answer. We also think given the practical constraints that are a necessary part of this kind of fieldwork, the investigation of the case involving multiple trips to Turkey to interview multiple witnesses was quite sound.

H. H. Jürgen Keil
School of Psychology, University of Tasmania
Hobart, Tasmania 7001, Australia

Jim B. Tucker
Division of Perceptual Studies
University of Virginia, Charlottesville, Virginia USA
jbt8n@virginia.edu

Reply to Tucker and Keil

I would like to thank them very much for the attention that Tucker and Keil have given to my reappraisal of the case Kemal Atasoy. Their reply is, in several points, very informative and satisfying. However, I have the impression that the authors may have considered my suggestions somewhat unnecessary to guarantee the authenticity of the case, therefore not considering it profitable to employ them. I believe that some of the measures that I have suggested, if adopted more often, could help us better understand the modus operandi of the phenomenon underlying CORTs (cases of the reincarnation type), serving thus not merely as a safeguard against fraud or other alternative naturalistic explanations. This way, the use of psychological tests, not only providing us a better guarantee against possible explanations based on children's suggestibility, could also help in explaining, for example, why some of the children who had a violent death in a previous life have phobias, birthmarks, or birth defects in their present life while others (also having had a violent death in a previous life) have not. Could it be that this former group has the same profile of people who display stigmata or of people who have "relived" traumatic experiences with the help of hypnosis or drugs and then developed skin conditions similar to the ones they had during the original experiences? More psychological tests in children could help to address these types of questions. Therefore, I hope researchers will employ them more frequently, not only as part of the replication process that science calls for but also as a tool for unraveling the modus operandi of CORTs.

Finally, I still consider that the use of micro cameras would overcome, if not all, at least many of the problems that researchers face today in registering interviews, when compared with traditional cameras. The investigation of CORTs have always used a methodology much akin to forensic science, and a tighter similarity to it could only make the investigation more robust.

Vitor Moura Visoni
Rua Professor Manoel Ferreira, 144, Apto 608
Rio de Janeiro 22451-030, Brazil

Weight of the Soul: 28 Grams?

Carlos Alvarado's historical writings always make interesting reading, not least his work on Duncan MacDougall's experiment on the loss of weight at death—"On Duncan MacDougall's Experiment on the Loss of Weight at Death," Letter to the Editor, *JSE, 23*(3), Fall 2009, pp. 343–348. MacDougall's piece of research also gave me a surprising series of experiences:

In 1971, I published a book in Swedish, which later appeared in several foreign editions. The purpose of the book was to give an overview of research in parapsychology at that time, with emphasis on the survival problem. Among many other research reports, I mentioned the experiment by Duncan MacDougall in 1907.

The German edition of my book was published in 1973 or 1974. At the book release, a journalist looked in the book and saw my notice about the MacDougall experiment. He did not read very carefully, so he cabled out the news that I had done this research. During the following years, I received dozens if not hundreds of letters from around the world asking for reprints or details of my experiment. I became undeservedly famous for having weighed the soul. A French magazine made a joke of it and published a fun picture of me with a big 28-gram weight on my head. As late as 2000, a man in Germany published somewhere that I had made this experiment in the hospital of Kristianstad in 1972. (Retrocognitive precognition? I did not move to Kristianstad until 1979.)

The English edition of my book was translated by a woman living in a small town in the United States. In 1974 she read in her local newspaper that I had weighed the soul. She was obviously the only person in town to know the facts, so she called the editor. With a sigh he said: "It was the only good piece of news that day."

Nils O. Jacobson
Kristianstad, Sweden
nilsjacob@gmail.com

References

Alvarado, C. S. (2009). On Duncan MacDougall's experiment on the loss of weight at death. *Journal of Scientific Exploration, 23,* 343–348.

Jacobson, N. O. (1974). *Life without Death? On Parapsychology, Mysticism, and the Question of Survival.* New York: Delacorte Press/Seymour Lawrence. (Original Swedish edition in 1971, *Liv efter döden?* Göteborg: Zindermans)

Human Weight Loss upon Death

Regarding "Rebuttal to Claimed Refutations of Duncan MacDougall's Experiment on Human Weight Change at the Moment of Death", by Masayoshi Ishida, *JSE 24*(1):5–39, I agree with MacDougall that the body loses weight on death.

The body is an aqueous system under pressure, and dissolved gases are lost when the heart quits pumping. Just consider that a one-liter bottle of beer will lose 2 grams of CO_2 when the pressure is removed by taking off the cap.

May your entropy ever increase,
Frank G. Pollard

OBITUARY

Rémy Chauvin
(1913–2009)

Rémy (André Joseph) Chauvin was a French biologist and entomologist, Professor Emeritus at the University Paris V–Sorbonne. His brilliant scientific career led him to direct three different laboratories, successively: the Laboratory of biology of the bee at the National Institute of Agronomic Research (INRA) in Bures-sur-Yvette, from 1948 to 1964; the laboratory of psycho–physiology in Strasbourg, from 1964 to 1968; and the Experimental ethnology laboratory of the École Pratique des Hautes Études (University of Paris, Mittainville), from 1969 on. He was an authority on the science of social insects and birds. Tributes from his colleagues praise his enthusiasm, his openness, and his dynamism, qualities that made him an exceptional research director (Arnold, 2010).

Author of more than 50 books (with translations in 11 languages including Japanese), he wrote on topics ranging from the academic textbook to the science fiction novel, to the popularization of science, sociology of science, and philosophical essays. Chauvin wrote at least 200 scientific publications and supervised Ph.D. candidates for more than 35 years.

JSE long-time readers may know that Rémy Chauvin was an early Council member of the SSE. "Scientific exploration" truly expresses the process that motivated his research in frontier areas. I will describe his important role in the development of parapsychology in France and his other adventures into unorthodox areas.

Parapsychological Contributions

Chauvin discovered the books of J. B. Rhine about 1955. His scientific openness encouraged him to repeat the ESP experiments with Zener cards in order to "see for himself." His experimental subjects were usually himself or his relatives, including many of his young family members. To improve his knowledge and his confidence in the methods of scientific parapsychology, he decided to do a summer internship at Rhine's Parapsychology Laboratory in 1959. He thought he was the first Frenchman to go there, but Christiane Vasse, a teacher from Amiens, had been a visiting scholar in the summer of 1958 (Vasse & Vasse, 1958). Chauvin returned to the Rhine Center at the end of 1969 with his engineer student Pierre Janin. His friendship with Rhine was steadfast, and unwarranted criticism against him could put Chauvin into a huff (Lignon, 2010).

From 1962 on, Chauvin was involved with the Institut Métapsychique International (IMI), a French parapsychological foundation active since 1919. There he directed the committee on comparative parapsychology where he regularly presented his parapsychological work with animals. In 1973, following the death of philosopher Gabriel Marcel, Chauvin agreed to succeed him in the position of Honorary President of IMI, which he retained until his death. He gave courses on methodologies in parapsychology that were developed by IMI for the general public from 1974 to 1976. From 1975 to 1977 he was the first president of the Fédération des Organismes de Recherche en Parapsychologie et Psychotronique, which was an attempt to structure the field of French parapsychology.

On the experimental side, he invented a random generator based on the radioactive decay of uranium (Chauvin & Genthon, 1965, 1967, Chauvin, 1968a). He did several ESP experiments with cards (Chauvin, 1959, 1961, Chauvin & Darchen, 1963) and PK experiments with dice (Chauvin, 1964). With such work, he was the best representative in France of the scientific parapsychology of Rhine. But he also made a very original contribution with the famous nutritionist Jean Mayer, which was recognized in 1970 with the McDougall Award, a distinction given by Duke University's Parapsychology Laboratory. By combining his knowledge of ethology with his interest in parapsychology (Chauvin, 1961–1962), he created a random device that administered electric shocks on one side or the other of a cage in which a mouse was placed. The question to be answered was, would the mouse position itself beforehand to avoid the shock more often than chance would allow? The test did not require the intervention of an experimenter during trials (Duval & Montredon, 1968a, 1968b, Chauvin, 1986, Montredon & Robinson, 1969). This protocol is consistent with behavioural models then in place, but without adherence to their ideologies. The fate of this experiment is surprising. Walter J. Levy, then director of the Institute of Parapsychology, successfully replicated the study. But he was found committing fraud in 1974. The device acquired a bad reputation and was no longer used (Chauvin, 2002). Chauvin published this work under the pseudonym of Pierre Duval. His bitterness at having to use a pseudonym was so strong that he authored, as Chauvin, an ironic preface distancing himself from the positions of "Pierre Duval", the alleged author of *Nos Pouvoirs Inconnus* (1963). His co-author, Jean Mayer, used the pseudonym Evelyne Montredon, which was the name of Chauvin's grandmother.

Chauvin has also contributed in a more indirect way to the development of parapsychology in France. He has advised and supported many students who wanted to work in parapsychology, such as Janin and René Peoc'h. When he learned that psychology students formed a group at the University of Paris X–Nanterre, a group that would become the Groupe d'Études et de Recherches en Parapsychologie (Evrard, 2010), he simply moved to a professorship of biology at that campus of the university. He remained there only one year because, after the social movement of May 1968, students started to publicly challenge their teachers. Chauvin, headstrong, had to give a course, although all the students in the auditorium turned their backs to him (Lignon, 2010).

From the mid-1970s on, Chauvin defended parapsychology publicly. He did this in the media, he wrote the preface to many books, and above all he wrote many popular books about parapsychology. After fully examining the case of Rhinean parapsychology and its critics (Duval, 1963), he encouraged readers to go beyond this paradigm (Chauvin, 1980a, translated into English in 1985; see Chauvin, 1985), and then focused on applications of psi (Chauvin, 1981). These books have had a significant impact as a trigger for the vocations of many researchers in the field. For all this activity, he was the first Frenchman to be awarded the Outstanding Career Award from the Parapsychological Association, in 2002.

Contributions to Related Areas

Chauvin was more an anomalist than a parapsychologist: His curiosity was not limited to controlled laboratory phenomena, but extended to all those phenomena that we can study through science. With the develoment of the movement of "fantastic realism", including the book *Le Matin des Magiciens* (Pauwels & Bergier, 1960) and the journal *Planète,* Chauvin became part of an invisible college of researchers, along with Jacques Vallée, Aimé Michel, Olivier Costa de Beauregard, and others. He participated in UFO studies (but he said he did not go beyond a literature review in his book *Le Retour des Magiciens,* Chauvin, 2002), in studies of mysterious history (Chauvin, 1980b), of instrumental transcommunication (Brune & Chauvin, 1993), and in a pioneering survey on highly gifted people (surdoués) (Chauvin, 1968b).

No idea was too extraordinary for him. However, his real motivation was to rebel against the dogmatic tendency of his contemporaries. He said that he started each day after having thrown away a misconception (Chauvin, 2002). Also, he strongly opposed scientist ideologies that acted as parasites to the practice of science itself, considering for example that the so-called Darwinians were neither Darwinians nor scientists because of their abuse of unverified or untestable hypotheses (Chauvin, 1997). He often said: "Scientism leaves the field open to charlatans" (Lignon, 2010). Science was for him the way to reveal the unknown behind appearances. A scientist embodying such a rigorous practice of subversion could only be an explorer, an anomalist.

This approach disturbed some of his fellow parapsychologists, who did not follow all of his arguments. When he supported the idea that communication with the Beyond was proven through instrumental transcommunication, he confessed that it was because of a weakness for needing proof, because although he was a scientist he had a deep faith in the Christian God. However, he expressed a naturalist compromise with this sentence: "If God and Satan exist, they use the laws of nature that I study" (Lignon, 2010). Intellectually free, modest, and a humanist, he became a standard bearer for half a century of exploratory science.

RENAUD EVRARD
Department of Psychology, University of Rouen
Mont Saint-Aignan 76821, France
home: Hagondange 57300, France
evrardrenaud@gmail.com

References

Arnold, G. (2010). *Le Pr Rémy Chauvin et les Abeilles.* http://www.sauvonslesabeilles.com/IMG/pdf/Remy_CHAUVIN__Dr_Gerard_ARNOLD_.pdf

Brune, Père F., & Chauvin, R. (1993). *En Direct de l'au-Delà. La Transcommunication Instrumentale, Réalité ou Utopie?* Paris: Robert Laffont.

Chauvin, R. (1959). Influence of the position of the subject in relation to the test cards upon ESP results. *Journal of Parapsychology, 23,* 57–66.

Chauvin, R. (1961). ESP and size of target symbols. *Journal of Parapsychology, 25,* 185–189.

Chauvin, R. (1961/62). Über die Möglichkeit parapsychischer Phänomene bei Tieren. *Zeitschrift für Parapsychologie und Grenzgebiete der Psychologie, 5,* 142–154.

Chauvin, R. (1964). A test of whether learning took place in a PK experiment with dice. *Journal of Parapsychology, 28,* 293.

Chauvin, R. (1968a). PK and radioactive disintegration. *Journal of Parapsychology, 32,* 56.

Chauvin, R. (1968b). *Les Surdoués.* Paris: Stock.

Chauvin, R. (1980a). *La Parapsychologie. Quand l'Irrationnel Rejoint la Science.* Paris, France: Hachette.

Chauvin, R. (1980b). *Les Secrets des Portulans ou les Cartes de l'Inconnu.* Paris: Empire.

Chauvin, R. (1981). *La Fonction psy.* Paris: Robert Laffont.

Chauvin, R. (1985). *Parapsychology: When the Irrational Joins Science.* Jefferson, North Carolina, USA: McFarland.

Chauvin, R. (1986). A PK experiment with mice. *Journal of the Society for Psychical Research, 53,* 348–351.

Chauvin, R. (1997). *Le Darwinisme ou la Fin d'un Mythe.* Monaco: Editions du Rocher.

Chauvin, R. (2002). *Le Retour des Magiciens.* Paris: JMG Editions.

Chauvin, R., & Darchen, R. (1963). Can clairvoyance be influenced by screens? *Journal of Parapsychology, 27*(1), 33–43.

Chauvin, R., & Genthon, J. P. (1965). Eine untersuchung über die moglichkeut psycho-kinetscher experiment mit uranium und geigerzalher. *Zeitschrift für Parapsychologie und Grenzgebiete der Psychologie, 8,* 140–147.

Chauvin, R., & Genthon, J. P. (1967). An investigation on the possibility of PK experiments with uranium and a Geiger counter. *Journal of Parapsychology, 31,* 168.

Duval, P. [Rémy Chauvin] (1963). *Nos Pouvoirs Inconnus.* Paris: Editions Planète.

Duval, P. [Rémy Chauvin], & Montredon, E. [Jean Mayer] (1968a). A PK experiment with mice. *Journal of Parapsychology, 32,* 153–166.

Duval, P. [Rémy Chauvin], & Montredon, E. [Jean Mayer] (1968b). Further psi experiments with mice. *Journal of Parapsychology, 32,* 260.

Evrard, R. (2010). French parapsychology after May 1968: A History of GERP. *Journal of Scientific Exploration 24*(2):283–294.

Lignon, Y. (2010). Rémy Chauvin (1913–2009). Public presentation (unpublished) and workshop given at the Institut Métapsychique International meeting; Paris, France; January 29, 2010.

Montredon, E., & Robinson, A. (1969). Further precognition work with mice. *Journal of Parapsychology, 33,* 162–163.

Pauwels, L., & Bergier, J. (1960). *Le Matin des Magiciens.* Paris: Gallimard.

Vasse, C., & Vasse, P. (1958). ESP tests with French first grade school children. *Journal of Parapsychology, 22,* 187–203.

BOOK REVIEWS

The Tunguska Mystery (Astronomer's Universe) by Vladimir Rubtsov. Springer (Heidelberg/London/New York), 2009. 318 pp. $29.95 (hardcover). ISBN 978-03877565730.

Hundreds of articles and dozens of books have been written on the Tunguska mystery, offering a variety of solutions to the nature of the phenomenon that occurred over central Siberia in the early morning hours of June 30, 1908. This book differs from others on the subject in key respects: It is interdisciplinary in nature, the author is a Ukranian schooled in Russia and thus able to present a detailed synthetic view of Russian research on the subject over the last century, he is technically competent yet has a broad background in the philosophy of science, and he has no apparent axe to grind, despite flirting with the extraterrestrial spaceship hypotheses as part of his admirable plea to take into account all possibilities. After a century of research on the subject, including 35 years of his own personal research, he believes a definitive answer has not yet been found to the cause of the explosion, but that we are coming ever closer to an answer if only the proper effort would be applied.

Any solution, he argues logically, must be based on the empirical evidence, not on theoretical calculations. What is certain from the numerous accounts is that a fiery object was seen entering the atmosphere accompanied by loud sounds but no smoky trail, and that the object exploded with a force up to 40 to 50 megatons, equivalent to 3,000 Hiroshima bombs. The explosion, heard more than 800 kilometers from the epicenter, leveled 2,100 square kilometers of landscape with millions of trees extending from the epicenter, where partially burned tree stumps were found. No primary crater or debris has ever been found, despite numerous expeditions to the area beginning with those of Leonid Kulik, a meteorite specialist from the Russian Academy of Sciences, in 1921 and 1927. In the 1930s the British astronomer Francis Whipple (not to be confused with the American comet expert Fred Whipple) suggested an exploding comet as the cause, and in 1946 the Russian engineer and science fiction writer Alexander Kazantsev (1906–2002) first hypothesized it was an exploding spaceship in his story "The Explosion." According to Rubtsov, one of the stranger phenomena not widely known is that the skies of Europe and Russia were anomalously illuminated for three nights preceding the explosion. More explicable are the observatories, including the Mt. Wilson Observatory, that reported a decrease in atmospheric transparency for several months after the event. Rubtsov also describes the results of Russian investigations showing that local geomagnetic effects were associated with the explosion. In addition, the explosion registered on seismic stations across Eurasia. Thus, as Rubtsov points out, we are left with three traces of the explosion, mechanical (felled trees), thermal (burnt tree remnants) and magnetic, along with some lesser clues that he examines in detail.

For the last 60 years, opinion in the former Soviet Union has been divided into two camps, those favoring natural versus those favoring artificial explanations such as the spaceship hypothesis. The latter has virtually no following in the West, where, following the work of Chris Chyba and others, an airburst of a stony meteorite is the leading interpretation. For some reason, Rubtsov does not mention Chyba's work, published in 1993. In any case, he is skeptical of the general meteorite interpretation, arguing (p. 252) that neither the cometary nor asteroidal hypotheses can explain the event, mainly because neither a crater nor debris have been found. However, in 1999 an Italian team of scientists led an investigation to Lake Cheko, some 8 kilometers north of the epicenter. Rubtsov gives only one sentence to the Italian expedition (p. 103), probably because its results were only announced in 2007 and published in 2008 as this book was in production. But the findings were spectacular: Seismic reflection profiles showed evidence of a dense, meter-sized rocky object at the center of the lake, which the scientists determined fills a space shaped like an inverted cone rather than the usual flat lakebed. They hypothesize this is a small fragment from the main colliding body, whether asteroidal or cometary. Another expedition returned to the site in 2009, and plans are under way to drill at the center of the lake—no small task in central Siberia.

If the cause of the Tunguska event was the entry into Earth's atmosphere of a Near Earth Object (NEO), it is incentive for more study of these objects, a plea that Rubtsov makes at the end of his volume. NASA (the U.S. National Aeronautics and Space Administration) has devoted considerable research to the problem, and maintains an NEO program at the Jet Propulsion Laboratory (JPL). With events like Tunguska in the background, many believe that funding needs to be increased on the search for NEOs and on possible deflection strategies. Indeed, to some visionaries (including Michael Griffin, the recent NASA Administrator) one of the motivations for spaceflight is the ability to remove a few representatives of homo sapiens from the home planet, in case of a catastrophic event that would cause a mass extinction and require starting life over with the slime of 3.8 billions years ago. Although a long shot, it would seem a small price to pay to hedge our bets, and it is not so far-fetched considering the recent confirmation that the Chicxulub crater in Mexico was caused by a large asteroid 65.5 million years ago, resulting in one of the three largest mass extinctions in the last 500 million years.

Moreover, we continue to have frequent, if small, reminders of the dangers to the Earth from space, most recently the object 2008 TC3, a meteoroid 2 to 5 meters in diameter that burned up over the Sudan on October 7, 2008, having been detected the prior day, resulting for the first time in an accurate impact prediction and a warning from JPL's NEO program. Unlike Tunguska, in this case 8.7 pounds of 280 meteoritic fragments were found, raising the question: Even if the Lake Cheko object is confirmed, why have no other fragments been found?

In the end, this book, well-written and meticulously referenced, is a laudable attempt to look at the available evidence while keeping an open mind. In many ways the Tunguska event is similar to the UFO debate, with too little evidence giving rise to too many speculative hypotheses. But in the Tunguska case, all agree there is physical evidence that the event actually occurred and was not just an aberration in the mind of the observer or a psychosocial phenomenon. With the Lake Cheko evidence, the intriguing century-long mystery of Tunguska seems finally to be yielding its secrets to science. Should that prove to be the case, it will reemphasize the importance of Carl Sagan's dictum that extraordinary claims require extraordinary evidence. And it raises the question: What are the limits of an open mind?

STEVEN J. DICK
21406 Clearfork Ct.
Ashburn, VA 20147

Witness to Roswell: Unmasking the Government's Biggest Cover-Up by Thomas J. Carey and Donald R. Schmitt. New Page Books, 2009. 318 pp. $16.99. ISBN 9781601630667.

There are hundreds or thousands of unexplained UFO sightings reported by people from all walks of life. If taken literally, these reports suggest that nonhuman intelligences (NHI) are traveling in machine-like craft through the atmosphere, landing and taking off from the ground and entering and leaving bodies of water. These reports are strongly suggestive, but they are not "hard evidence," in the sense of a piece of unknown material ("unknownium") or an alien creature's body.

And then there is Roswell.

Anyone who hasn't heard of Roswell (New Mexico) has been living on another planet or in a cave or underwater or in the air, suspended by a Mogul balloon, for the last three decades. Quite possibly the longest running UFO investigation of all, the Roswell case began in July, 1947, took a break for about 30 years, resumed in late 1978, and continues to this very day, with the book reviewed here as the most recent summary of what is known. Although the original version of this book, published in 2007, contained a lot of information from witnesses, I strongly recommend that the reader obtain the 2009 version which contains witness information as recent as late 2008.

The Roswell crash first became known to the general public shortly after noon, July 8, 1947, when the Roswell Army Air Force Base issued a press release that said the Army had acquired a flying saucer. A few hours later, General Ramey at Fort Worth, Texas, under pressure from the Pentagon to "get the press off our backs," announced that it was only a weather balloon. Years later, in 1994, the Air Force admitted that it wasn't simply a weather balloon. Instead, the Air Force claimed that it was a "Mogul balloon." Further, in 1997, the Air Force announced that any bodies associated with the Roswell crash were actually "crash dummies," anthropomorphic, six-foot-tall mannequins that were designed to test survival techniques for pilots ejecting from high-flying aircraft and manned satellites. (This is what I call the "dummy drop" theory of Roswell.) One problem: The dummies weren't dropped until 1953, and they were not dropped in the places where Roswell wreckage was found.

People who have not followed the most recent Roswell research will probably be surprised, or maybe even astounded, at the mass of testimonial evidence. No one now doubts that *something* unusual crashed in the desert outside Roswell. The question now is, what was it? The United States Government and skeptics in general have settled upon the "Mogul Balloon Train Hypothesis" (MBTH). Other proposed explanations have been superseded by the MBTH. The reader of this book will have plenty of information on which to base his/her conclusion as to whether or not the MBTH could possibly be correct. As for me, when I learned of the testimony of Jesse Marcel, Jr., the Roswell base intelligence agent, I decided that any explanation based on ordinary balloon materials, including the MBTH, was insufficient. Skeptics argue that the Mogul project was highly classified and Marcel had no "need to know" about the project so he wouldn't be aware of the large amount of ordinary balloon material associated with a crashed Mogul balloon. However, the key here is "ordinary." The Mogul balloon device consisted of many ordinary balloons plus an ordinary radar reflector. It is very difficult to imagine that Marcel, trained in radar techniques, would fail to recognize ordinary balloon material and ordinary radar reflector

Book Reviews

material. And, if Marcel were to fail to recognize the material, surely other people at the base would have.

Of particular interest in this regard is Captain Sheridan Cavitt, in charge of counterintelligence at the base. According to Marcel, Cavitt accompanied him to the crash scene along with the farmer who had found the material, Mac Brazel. When Cavitt was first located in the 1980s by investigator Bill Moore, Cavitt would not admit to anything. However, when interrogated by the Air Force as part of the 1994 Air Force investigation of Roswell, Cavitt admitted that he had gone with Marcel, and, furthermore, he claimed that he recognized the material immediately as a weather balloon (not a Mogul Balloon). So, why didn't he tell Marcel? Cavitt's testimony is about a solid as a Swiss cheese. You can read about Cavitt's testimony in the book, but for a more detailed analysis read "Cavitt Emptor" at http://www.brumac.8k.com/CavittEmptor.html.

One of the witness stories in the book is that of Dr. LeJeune Foster, an anthropologist. As I read her story I was reminded of what I was told by Isabel Davis, the secretary of NICAP (National Investigations Committee on Aerial Phenomena) in the 1960s and cofounder of the Fund for UFO Research in 1979. One day in the 1970s (I cannot now recall exactly when this was) she told me the following story which I will recite to the best of my ability after hearing it more than 30 years ago. Isabel was "into" UFO research in the late 1950s at a time when she met a lady scientist (she would not tell me the name) who had a high level security clearance and who said she had been asked to do an anatomical study of a "creature." According to Isabel, this lady had been taken by the government to a secure laboratory in the Washington, D.C., area. Before entering the room with the creature she had to take off all her clothes and put on some special laboratory clothing. She was told to walk through several doors, such as would form an airlock, and was then inside and alone. She found the body of the creature and all the necessary instruments and devices she needed to make her study of the body. It was clearly not human, she told Isabel. She wrote her impressions and conclusions in a notebook that was in the room. When she was finished she took nothing with her and exited through the multiple doors. She was told to keep quiet about what she had just seen and done and was then transported home.

Clearly, by telling Isabel, this woman had violated the security requirement that she not tell anyone. I asked Isabel, why did this lady dare to say anything about it? Isabel said she asked the lady that same question. The lady said to Isabel something like this: It is only a story, and I have no hard evidence that this happened; furthermore, if confronted I will deny that it happened.

Over the years I often wondered who that lady scientist might have been, and then I learned of the story of LeJeune Foster. Could she have been the lady scientist? Unfortunately, Isabel Davis died many years before Foster's story was uncovered so I had no way of confirming the identification.

This is just one of numerous witness stories in this book. Anyone seriously interested in the Roswell research should read this book to find out just how much testimony is available. Lots of people had a "piece of the action" and when these pieces are put together the stories point to one thing: An object/craft crashed near Roswell and alien creatures were found and the government has tried to cover it up.

BRUCE MACCABEE
brumac@compuserve.com

The Roswell Legacy: The Untold Story of the First Military Officer at the 1947 Crash Site by Jesse Marcel, Jr., and Linda Marcel. New Page Books (Franklin Lakes, NJ). 2008. $14.99 (paperback). ISBN 978-1601630261.

It may not be an exaggeration to say that somewhere in the world at every hour, details of the crashed object at Roswell, New Mexico, are discussed, broadcast, or printed in one form or another. It is also an unusual story in that even though it was national news when it was first reported, it never really became popular until more than thirty years later. At this time, witnesses began to reveal a variety of accounts of their personal knowledge of a "flying disc," as the press called it early on, that the Army Air Force (AAF) retrieved and displayed to reporters on July 8, 1947. Within a matter of hours after that exposure, the "disc" was relegated to being merely a weather balloon and the story passed into obscurity.

The first person known to have contradicted the official dismissal of Roswell was former Major Jesse Marcel, Sr., the base intelligence officer who retrieved debris for the AAF after being notified by a witness, civilian rancher "Mac" Brazel. Marcel surfaced after a TV station manager in Baton Rouge, Louisiana, informed UFO researcher Stanton Friedman in 1978 that he heard Marcel tell the story of Roswell being more than just a balloon. Friedman contacted Marcel and helped to initiate modern publicity about the alleged extraterrestrial nature of the event. This in turn led to others coming forth, creating an ever-enlarging Roswell snowball.

After the spate of books and articles on Roswell since 1980 by numerous authors, *The Roswell Legacy* comes from the family of the original witness who had essentially revived a dead controversy. Jesse Marcel, Jr., the son of Major Marcel, is a Veterans Administration doctor in Montana who has had a lengthy career in the military, as did his father. Before passing away in 1986, the father struck an agreement with his son to make the true story available to the public.

This meant of course the story as the Marcels had seen it. Yes, the Marcels, because Jesse Marcel, Jr., wasn't just the author of his father's involvement but was an active participant. He and his mother were shown a box of debris gathered by his father and brought home after the first field trip to inspect the crash site at the Foster Ranch. The authors, Jesse Marcel, Jr. (hereon referred to as "Jesse Jr.") and his wife Linda, go into considerable detail about the debris. They described plastic-like material, small "I" beams, and an abundance of foil-like material. Jesse Jr. thought the material was "interesting", but at the same time felt (p. 53) "I didn't really understand what all the excitement was about. It surely did not seem to be anything worth getting up in the middle of the night to see." This may have been due to the fact that Jesse Jr. was only 11 years old at the time and flying saucers were a relatively new topic.

Questions could certainly be raised about the accuracy of an 11-year-old's recollections, especially in light of how his father was seen to be very excited and told Jesse Jr. that these were parts from a "flying saucer" (p. 53). In that this was perhaps the most important part of the book with respect to demonstrating whether or not debris of a possible spacecraft was recovered, the question in my mind was whether or not the story told by the Marcels was convincing.

Nothing resembling computer electronics, or even machine parts, was evident in the debris. The bulk of the pile seemed to be foil-like sheeting with plastic "I" beams displaying nondescript symbols. There was no context for how the debris may have fit together. The pieces were part of a structure of some sort, and in the mind of Jesse Marcel it was a flying saucer. Seeing the description and photos of Roswell debris in various accounts, I found it difficult to think that a ship from another planet could be made from such flimsy building material. Maybe I just don't know a spacecraft's engineering structure the way I should! But looking back upon contemporary accounts of the original story, W. W. Brazel told reporters that what he found consisted of "large numbers of pieces of paper covered with a foil-like substance and pieced together with small sticks, much like a kite" (*Las Vegas* [New Mexico] *Daily Optic*, July 9, 1947). The largest piece was said to be three feet across, and the debris, including small pieces of gray rubber, were scattered over a 200-yard area.

Once again, this description, not unlike how Jesse Jr. described his view of the debris he saw, doesn't give the impression at all of a technology from outer space. One exception offered in the book though is Jesse Marcel's explaining how one of the men at the base tried to hit a piece of the debris with a sledgehammer, only to see the tool bounce off of the piece without doing any damage whatsoever. He added that he himself tried to crumple another piece in his hand which, when released, returned to its original shape with no evidence of crumpling. Unfortunately, Jesse Jr. did not see any of this happen with the debris in their home.

If there were smooth, undamaged pieces of the Roswell debris, as described by Jesse Marcel, they surely weren't evident in the scraps that were photographed by J. Bond Johnson, a *Fort Worth* [Texas] *Star-Telegram* representative who was present at a public display of Roswell wreckage in the office of the 8th Air Force commanding general, Roger Ramey, on July 8th (1947). Two photos by Johnson in the book give cause to wonder what became of the mystery smooth debris often cited in Roswell accounts. The authors give a rather confused explanation for this, saying that Johnson saw only a small portion of the "actual" debris and that he was allowed to observe wrapped-up "real" debris from a distance so his camera wouldn't get good detail of it. They wondered, "Does this mean that there was a mix of genuine debris with debris from a weather balloon?" (p. 69). Answering their own question, the authors say, "This is indeed what happened", not seemingly concerned with why real debris would be mixed in with unreal debris at all if a grand coverup of the truth were under way.

More confusing still is that after the first contact with Jesse Marcel by Stanton Friedman, Friedman's partner at the time, William Moore, interviewed Marcel, with the interview ending up in the book Moore wrote with Charles Berlitz in 1980, *The Roswell Incident*. We see this on page 68 regarding the Ramey office debris:

> General Ramey allowed some members of the press in to take a picture of the stuff. They took one picture of me on the floor holding up some of the less-interesting metallic debris. The press people were allowed to photograph this, but were not allowed into the room to touch it. The stuff in that one photo was pieces of the actual stuff we had found. It was not a staged photo. (Marcel quoted in *The Roswell Incident* by Charles Berlitz and William L. Moore, Grosset Dunlap, 1980)

He went on to say that this wreckage was cleaned out and replaced with substitute wreckage for other photos taken, while the real debris was sent out to Wright Field. Looking at the photos of Marcel with the debris and others with the debris in the Marcels' book, it didn't seem that the pieces were different in the respective shots.

There is a brief discussion of a strong odor of something burned from the pieces in the Marcel home. This may be a reference to the decayed rubber fragments that had been in the desert sun for an extended period of time. Such pieces were also present in General Ramey's office. The authors argue for the debris on Ramey's floor being different from what Jesse Jr. saw partly because he recalled no odor from the pieces gathered by his father and shown to his wife and son. Rereading the book's description of the pieces, the authors make no reference to the decayed rubber fragments, though his father must have seen them and picked them up since they were mentioned by Mac Brazel in his initial discovery of the pieces on the ranch. A simple explanation is that Jesse Marcel did not include the rubber in the wreckage brought home because they did smell. A coverup conspiracy can't be built on such an omission.

It is pretty clear that flying saucers were on the mind of Jesse Marcel from the first moment of his viewing of the site. The authors tell us that he made a number of statements early on to this effect, referring to it as an "unearthly craft" and a "flying saucer" and suggesting that the Marcel family were among the "first humans to see it." But the authors hint at a bigger picture. Jesse Jr. observed (p. 61),

> It seems he had seen other things that convinced him that this was not human manufacture. I didn't know what made him so strong in his beliefs, but because I had seen some pretty unusual features in the debris myself, and I trusted my father's expertise, it didn't take much to convince me that he was right.

This suggests the realization that the debris by itself was not convincing enough as evidence of a flying saucer, but instead it had to be tied to other accounts. To be that convincing, some of the tie-in stories would have to include the rumors of large saucer wreckage and bodies of aliens not evident in the original reporting in 1947. It is an odd fact that of the dozens of reports of observations of bodies and vast amounts of debris, the first documented witnesses out to the scene reported seeing neither. Why wouldn't the base intelligence officer know about alien bodies when a civilian Roswell mortician, Glenn Dennis, allegedly did? Or why, when wreckage and bodies were supposed to have been packed and flown out of Roswell, did the pilot and crew know about this but Jesse Marcel didn't? These are bothersome questions that the book doesn't clarify.

More bothersome still is this. As the base intelligence officer, Jesse Marcel was in his way responsible for preserving top security for a unit that was the only atomic bomb group in the country. He knew the ropes on how to deal with secrecy matters for a surprise development, such as the crash of a flying saucer. The book is emphatic that he believed this was a flying saucer. But instead of following security procedures and taking the wreckage immediately to a secure location, he brought it home to his wife and son first! Supposed pieces of an alien spacecraft was spread around the Marcel kitchen and handled with bare hands with no concern for any biological or radiological contamination. Then every last piece was gathered back together with the exception of small fragments on the kitchen floor that were eventually swept away with a broom and never kept or seen again. Marcel was conscious enough of proper procedure, having gone to intelligence school, to return all the material to the base for reasons of national security, but security had already been breeched in bringing the classified materials to a private residence.

The rest of the book contains the usual defense of Roswell-as-real using a good deal of what might be considered stretched information. For example, even though he didn't see this in the debris, Jesse Jr. cited the son of Mac Brazel as having seen strands of what looked like fiber in the portion of the debris that he saw. Shining a light on one end of the line revealed that the light appeared at the other end. Linda Marcel said Jesse Marcel told her, though he apparently didn't tell his son, that he saw the fine strands as well. It was inferred through third party information that the strands were a form of fiber optics, impressing technology onto the pile of debris by the inference. Of course the source of this claim did so more than thirty years after the event and no one saved a single strand of this material. Nothing can be demonstrated. A point well stretched.

The Ramey memo controversy is cited as supporting the mystery. The "memo" is a piece of paper held by Roger Ramey in one of the J. Bond Johnson photographs of the debris that appeared in the newspapers on July 9. The paper is at the edge of resolution in the photo and many conflicting attempts have been made to decipher it. These attempts have been reported in *JSE* (see *Journal of Scientific Exploration,* 16(1), Spring 2002, 45–66). Jesse Jr. seems to accept one of the decipherments discussing "victims" as a preferred reading, something by no means assured as an interpretation. It is not even embraced as a consensus among the various analyses of the image. In fact, there is evidence in the photograph suggesting that the "memo" is simply a teletype newswire from the offices of the *Fort Worth Star-Telegram.* Another point well stretched.

A full chapter is spent on a mysterious government official contacting Jesse Jr. to lend support to the story. Part of the support was stating that the Whitley Strieber novel *Majestic* was not fiction. *Majestic* deals in large part with endorsing the notorious "MJ-12" documents hoax of the 1980s, a story that has been thoroughly exposed as specious. With support like this, who needs debunkers? Stanton Friedman, the author of the book's Foreword, remains one of the last prominent defenders of MJ-12 being authentic. Jesse's contact sounds more like an enthusiast of government UFO secrecy than one knowledgeable on inside saucer information. The authors though do not spend much time on MJ-12 beyond this.

Considering the plethora of information on the Roswell crash over the past thirty years, readers are now looking for not just another Roswell book. They want information close to the story, meaning from those who were directly involved in it. And it has to be done quickly, since the pool of those directly involved grows ever smaller with the passage of time. *The Roswell Legacy* provides this perspective, so as such it should be read more so than the average Roswell volume. It sets out to do two things. It is a chronicle of Major Jesse Marcel's career and involvement in the saucer story, and it is an advocacy that a spacecraft was wrecked in the desert. It succeeds at both explanation and advocacy. But while it helps the reader to understand who Jesse Marcel was and the details of his version of Roswell, it doesn't prove extraterrestrials landed here nor does it mitigate the controversy. There are far too many issues that remain problematical.

BARRY GREENWOOD
uhrhistory@comcast.net

Art, Life and UFOs: A Memoir by Budd Hopkins. Anomalist Books (San Antonio, TX), 2009. 438 pp. $19.95 (paperback). ISBN 9781933665412.

Art, Life and UFOs is Budd Hopkins' autobiography. The author is probably best-known to readers of *JSE* as the quintessential proponent for the reality of alien abductions. He is the person most responsible for bringing worldwide attention to the phenomenon, his seminal work on the subject having built the foundation upon which the alien abduction experience is still understood. But Hopkins' first passion was as an artist in the Abstract Expressionist movement during the 50s and 60s, and this side of him continues to be as much a part of his personal sense of identity as his work with abductees. In this candid, fascinating, intimate, and often touching memoir, Hopkins chronicles the life and times of the man, the artist, and the abduction investigator. The reader will come away from *Art, Life and UFOs* feeling he knows Budd Hopkins as if he has been a lifelong friend.

As an artist, Hopkins hobnobbed with such contemporaries as Franz Kline, Willem De Kooning, Mark Rothko, Robert Motherwell, and Jackson Pollock, all the while developing his own artistic style in painting and sculpture. His works have earned considerable recognition during his career, and have been displayed at such museums as the Guggenheim, the Hirshhorn, the Museum of Modern Art, and the Whitney. But Hopkins can also paint great pictures with his words, and the tales he tells of life during the "Cedar Bar" years in New York City (named for the Cedar Street Tavern, an artist hangout in Greenwich Village) provide vivid imagery of the art world and its personalities during this period.

Hopkins reflects on his complex personal history with considerable introspection, weaving self-analysis throughout the tapestry of his life, the evolution of his art, and his emergence as the pre-eminent investigator of alien abduction cases. For example, an early battle with polio imposes restrictions on a normal childhood, nurturing a creative drive that serves a "need to invent my own private symbolic world", a world which "rely[ed] more and more on the inventive richness of my inner life". A childhood wonderment with nature portends a lifelong curiosity, a "need to know, to tell the false from the true", and to tell others the truths he has learned. As witness to the cruel and compassionless behavior of his boarding school classmates, Hopkins becomes a sensitive, caring supporter of tormented souls, a person who "befriends the untouchables".

These traits can be seen as essential to Hopkins' success in later years uncovering the abduction phenomenon. But to his detractors, these same traits are credited as fundamental to his creating an abduction belief system that has no actual basis in fact. For example, Philip Klass (the infamous debunker), noting that "fantasies of the mind are the source of most modern art and ... Hopkins' trade", held him responsible like no other for (as Klass saw it) the damage done to experiencers from the false beliefs instilled in them by abduction investigators. He called the abductees' affliction the "The Hopkins Syndrome".

Klass' opinions notwithstanding, Hopkins himself takes pride in the fact that "the discoveries I've made and published over the past 27 years collectively provide the foundation upon which abduction research has been built". We cannot know how history will ultimately treat this legacy, but Hopkins' assessment of his impact is unassailable, not just with regard to abductions but for ufology per se. Indeed, abductions have surpassed UFO sightings and Roswell as the most familiar and studied facet of ufology. On the down side, this has given ufology's debunkers the latest cause célèbre around which they cheerfully rally. On the up side, Hopkins' abduction scenario offers a raison d'être for the UFO presence, and answers the proverbial question of why they do not "land on the White House Lawn" (as he points out, the desire to keep such a nefarious operation secret is an understandable rationale for "the covert nature of the UFO phenomenon").

Paradoxically, in another of his books (*Witnessed*) Hopkins argues for the functional equivalent of the White House landing—in a case tantamount to an abduction on the United Nations Plaza. Of course, reports of apparently surreptitious behaviors conflicting with apparently ostentatious activity have long been part of UFO lore, and remain a puzzle for which ufology's proponents still have some explaining to do. Hopkins seems to think that the events he chronicles in *Witnessed* were designed as a deliberate warning to the authorities of just what the aliens are capable of (while remaining clandestine in most other respects). The more conventional wisdom holds that UFO activity reflects an orchestrated plan to gradually reveal alien presence to the world, and to make the world receptive to that revelation. The reader can decide if either attribution fits in the context of UFOs' rather flamboyant arrival on the scene in the 1940s, their unabated presence in our skies and (as often reported) on our military's sensing devices for at least the last 60 years, and/ or the apparent absence during this period of an increasing acceptance of UFO reality (in fact, if anything there may be a decreasing trend. A 1966 Gallup poll found that 48% of respondents felt UFOs "are something real [as opposed to] just people's imagination". This figure was 49% in Gallop's 1987 survey. According to a 2008 Harris Poll, only 36% surveyed "believe in UFOs").

Whatever dynamic might account for the reported behavior of UFOs, Hopkins' initiation into issues ufological began as a consequence of his own UFO sighting of a "daylight disc" in 1964. This Cape Cod event involved a multiple-witness observation of an apparently metallic lens-shaped object exhibiting aerial maneuvers not possible by conventional craft. The experience (and his innate curiosity) led Hopkins to plunge into the UFO literature and, as a consequence, realize that "these reports constituted a mystery I could not leave alone".

A second pivotal event in Hopkins' eventual emergence as abduction investigator came from an account told to him in 1975 by the proprietor of his neighborhood liquor store. He regaled Hopkins with a story of a close encounter with a 30-foot craft hovering just off the ground in North Hudson Park, North Bergen, New Jersey. As he was watching this object in amazement, a group of small humanoids descended from the UFO and proceeded to carry out an operation that appeared to be one of soil sample collection. The case was a classic "close encounter of the third kind", including landing traces, corroborating testimony, supporting ancillary evidence, etc. Hopkins' investigation, which he published in the *Village Voice*, received considerable media exposure at the time, and was later described in detail in his first book, *Missing Time*. In the book currently under review, Hopkins reveals a new wrinkle that, as with many UFO cases, just adds to its strangeness. He describes how seven years prior to the reported event, he visited the building across from the park where it took place—in order to deliver a painting commissioned by one of its tenants. It was the only apartment building in New Jersey he had ever been in until he returned to work on this investigation. To compound that synchronicity, the building's doorman, a witness who provided testimony corroborating the liquor store proprietor, remembered Hopkins from his visit years earlier. Hopkins describes his reaction to learning this:

> If the North Hudson Park UFO landing case had been unnerving before, the shock I felt now had doubled. What were the odds that I would have known both witnesses to this bizarre incident long before it took place? Neither had ever met the other, but I knew them both. In advance! So upsetting to me was this micro-coincidence, that I have never before made it public.

As his own sighting opened Hopkins' mind to the general idea of UFO visitations, this close encounter case got him thinking more and more about the UFO occupants, and what their agenda might be. Prominent abduction cases like those of Betty and Barney Hill, Calvin Parker and Charles Hickson, and Travis Walton began to shape his interest in abductions, as did a possible abduction case involving a personal acquaintance. Before long Hopkins was investigating abduction cases in earnest, embarking on a path that would make him the foremost investigator of alien abduction of humans in our time. His work showed that abduction experiencers report events that have many features in common (discloser: Hopkins and I, along with Don Donderi, have shared credits on several conference papers analyzing some of these commonalities), that the events they report often involve corroborating testimony from multiple witnesses, and that the reporters of such encounters seem reliable and sincere. As incredible as the implications of his investigations seemed to be, for Hopkins the cumulative evidence was so compelling that he "no longer had the luxury of *dis*belief" (his emphasis).

These initial forays into abduction research established the essential framework that he would so effectively promulgate in the ensuing decades:

- Alien control of abductee motor function, facilitating their being taken against their will
- Missing time: Unaccounted-for periods following intact memories of an initial close encounter (or sometimes just a vague sense that "something happened")
- Hypnotic regression as a tool for retrieving memories from these periods of missing time
- The revelation of repeated abductions throughout an experiencer's life
- Screen memories masking actual alien encounters
- Paralysis and psychological manipulation (fear, etc.) of the abductee
- Examinations and other physical procedures conducted during the abduction

The ultimate centerpiece of the abduction scenario, the hybridization program, would evolve later in Hopkins' thinking. This first took shape during his investigation of the "Kathie Davis" case, chronicled in his second book, *Intruders: The Incredible Visitations at Copley Woods*. The experiences reported by Davis left Hopkins with

> only one conclusion [to] draw. Apparently the central purpose of the systematic alien program of human abductions is the creation of genetically altered beings—part alien, part human hybrids.

As his investigations and writings gained increasing attention, Hopkins' newfound celebrity brought him into contact with such media personalities as Oprah Winfrey, Larry King, Bryant Gumble, Regis Philbin, and Matt Laurer, and set the stage for a notable telephone interview with Walter Cronkite. He also gets to know other luminaries making up the ufology/abduction landscape: J. Allen Hynek, David Jacobs, John Mack, Carl Sagan, Whitley Strieber. A vignette about a meeting with Shirley MacLaine (herself a proponent of alien encounters) is particularly intriguing.

All in all, Hopkins has enjoyed a very interesting life, while making significant contributions to art, popular culture, the contemporary psyche, and our awareness of one of the most provocative phenomena of this or any other time. For those unfamiliar with his story, or just curious about his personal take on it (this is the only true autobiography by any major figure in contemporary ufology), *Art, Life and UFOs* is highly recommended.

STUART APPELLE
Dean, School of Science and Mathematics
The College at Brockport, State University of New York
Brockport, New York, 14420
sappelle@brockport.edu

Bigfoot: The Life and Times of a Legend by Joshua B. Buhs. University of Chicago Press, 2009. 304 pp. $29.95 (hardcover). ISBN 9780226079790.

Anatomy of a Beast: Obsession and Myth on the Trail of Bigfoot by Michael McLeod. University of California Press, 2009. 238 pp. $24.95 (hardcover). ISBN 9780520255715.

> Joshua Buhs explains that *Bigfoot: The Life and Times of a Legend*
>
> picks up where the folklorists stopped, trying to understand how Bigfoot became prominent in American culture, why some people believed the creature existed, the function that such belief served, and how the debate over the existence of wildmen fit into twentieth-century American culture.

As early as page four, he establishes his perception of bigfoot as nonexistent in the caption of a wildman drawing from 1490, noting that "Bigfoot is a modern example of a well-known mythological archetype." Additional examples from wildman literature develop his view of the sasquatch or bigfoot as exclusively mythical: "First, this book shows how the modern myth of Bigfoot emerged out of, and diverged from, traditional wildman tales.... Second, this book connects these modern tales of wildmen to concerns over the maturation of mass culture and consumerism." Buhs explains his preoccupation with the mass media as follows: "Bigfoot was born of the mass media, spread on the mass media, and its vitality came from the fear of mass media and consumerism."

At one point in the book, Buhs appears to hedge his bet regarding the existence of bigfoot, conceding that "Indeed, it's not impossible that an actual Wildman may someday be caught." In the last paragraph, however, he categorically concludes that "...Bigfoot did not exist . . . (the skeptics were right.)"

His attempts to lead readers to this conclusion consist of descriptions of the best-publicized events and activities related to bigfoot investigation during the past half century: conferences, expeditions, and database formation, focusing on the personalities involved. It is-well documented and will be of interest to those curious as to why the subject has been ignored by most scientists in the larger scientific community.

Not surprisingly, Buhs is most aware of the charismatic and outspoken proponents of bigfoot, those investigators with the greatest media presence. Consequently, the picture painted is not flattering to bigfoot investigation. On the other hand, the book provides an interesting and enlightening read, bringing to light the background of many episodes in bigfoot investigation. Examples of mismanaged expeditions, personality clashes, inappropriate—even rude—behavior at conferences, and errors in methodology are described at length.

A potential problem arises in that Buhs appears to have accepted the foibles, inappropriate behavior, and character flaws of some investigators as not just entertaining, but also as a basis for his conclusion regarding the nonexistence of bigfoot. Undiscerning readers might do the same. For example, a *Publishers Weekly* review of *Bigfoot: The Life and Times of a Legend* (2009) included the following praise: "Buhs is at his amused best when following the exploits of Bigfoot's human handlers, the colorful band of true believers, hoaxers, and pseudo-documentarists who constructed this greatest of all shaggy dog stories." Perhaps the reviewer's previously held views were affirmed by this book, or possibly they result in part from the descriptions of how bigfoot has been investigated and by whom.

Although Buhs cannot be held responsible for comments such as these, they show that, while being enlightened and entertained, at least one reviewer got the message that investigators collecting and documenting sasquatch tracks and eyewitness descriptions are *pseudo*-documentarists. Serious investigators may chafe at this designation, but they may be even more disconcerted by being lumped in with hoaxers—those individuals who have so effectively contributed to the taboo nature of sasquatch research, negating its validity as a subject of scientific study and repelling scientists who are already skeptical.

*

If Buhs found the efforts of amateur investigators in the past 60 years to be merely amusing, Michael McLeod appears to have found them positively offensive. In *Anatomy of a Beast: Obsession and Myth on the Trail of Bigfoot*, McLeod explains that "Bigfoot is more than just a silly slice of history. The beast's appearance on the national scene marked an important milestone: the first widely popularized example of pseudoscience in American culture." Indeed, "The increasingly common use of pseudoscience—junk science—has transformed public debate, as reflected in the anti-intellectualism now sweeping the country. This book makes the case that it all began with Bigfoot."

Emphasizing his disregard for the acceptance of bigfoot as extant, McLeod asks rhetorically: "If people can delude themselves into believing in the existence of an eight-foot tall ape man, what on earth might they be thinking about truly important matters?"

While Buhs' writing hints of smugness as he describes expeditions gone awry or his perception of gullibility in an investigator, McLeod leaves no such doubt in his descriptions of investigators, their goals, and activities. Buhs reserves a tiny window of doubt regarding bigfoot's existence: "I still don't think that bigfoot exists—indeed writing this book actually gave substance to what was before only a vague kind of skepticism." McLeod, however, gives the impression that people who do not agree with him are themselves deluded. Since neither author appears to consider the possibility that there could be readers who do not share their viewpoint, they consequently express their confident assertions as if "preaching to the choir."

*

A quote from *The Critical Historian* by George S. R. Kitson Clark may explain their tone of certainty:

> When a conflict is over, historians are too inclined to take the case for one side and all its partisan stories straight into the canon of history without looking at the evidence, or trying to find out what the other side may have had to say. (Clark, 1967)

Both Joshua Buhs and Michael McLeod appear to have concluded that any conflict regarding whether or not bigfoot is an extant mammal is over. And, not surprisingly, they perceive the supposed conflict as over, in favor of bigfoot as merely a cultural phenomenon, particularly as a hoax.

But any supposed "conflict" is far from over, and—in the eyes of a small minority of scientists—has yet to take place. What Buhs and McLeod have written about is far from representative of what "the other side" has to say. Both authors largely ignore the existence of the small, persistent cohort of scientists who, along with an equally dedicated and much larger group of amateur investigators, are quietly at work collecting and archiving evidence which only occasionally attracts the attention of scientific colleagues or the media. Although Buhs lists several scientists, he does not discuss their work in any detail, nor does he mention ongoing research. The reader is left to wonder if he is aware of it and just chooses not to discuss it. Although such scientists might be described by McLeod as having "deluded" themselves, there is a growing number of scientists with an undisclosed interest in the subject, some of whom quietly support sasquatch research.

If read carefully and critically, these books reveal much of interest. They elucidate in considerable detail why some people may have dismissed the possible existence of bigfoot on the basis of how it has been investigated and by whom. In fact, both of these authors imply that such reasoning has influenced them to "throw out the baby with the bathwater." Fallacious reasoning such as this is, however, the very hallmark of pseudoscience. Consequently, it may be especially ironical that McLeod identifies the study of bigfoot as not just an example of pseudoscience, but as the iconic beginning of a pseudoscientific trend.

JOHN BINDERNAGEL
Author, North America's Great Ape: The Sasquatch: A Wildlife Biologist Looks at North America's Most Misunderstood Large Mammal (1998)
johnb@island.net

References

Clark, George S. R. Kitson (1967). *The Critical Historian*. New York: Basic Books. pp. 51–52.
(2009). Review of *Bigfoot: The Life and Times of a Legend* by Joshua Buhs (Chicago: Chicago University Press). *Publishers Weekly*, March 2, 2009.

Philosophy in the Flesh: The Embodied Mind and Its Challenge to Western Thought by George Lakoff and Mark Johnson. Basic Books (New York), 1999. 624 pp. $24.95 (paperback). ISBN 9780465056743.

This *magnum opus* of so-called "second-generation" cognitive science begins with the bold declaration that "three major findings" have brought to an end "[m]ore than two millennia of a priori philosophical speculation" (p. 3):

The mind is inherently embodied.
Thought is mostly unconscious.
Abstract concepts are largely metaphorical.

Fleshing out innumerable implications of these findings, the authors advance a new approach to philosophy, dubbed "embodied realism" (pp. 95, 74–93). While the truth of these findings may be unassailable, grounded as they are in empirical fact, most of the book defends the thesis that embodied realism is "at odds with" *all* past approaches (p. 548), requiring philosophy to begin anew. This polemic fails for two main reasons.

First, the authors' many extreme claims cannot be substantiated by the (often impressive) empirical evidence presented. For example, proving that "reason is not, *in any way*, a transcendent feature of the universe" (p. 4), simply because we inevitably use embodied metaphors in reasoning, would require adopting the very disembodied standpoint that is being denied. In place of such dogmatic declarations, humbly confessing ignorance regarding what (if anything) transcends the embodied mind, driving us to employ metaphors, would be more defensible. Apparently unaware of this fundamental lesson of Kant's critique of reason, the authors focus their most vehement attacks on Kant; denying validity to this (and every other) classical philosophical system proves only their weak grasp of the philosophical tradition they aim to supplant.

The second reason the authors fail to reach their stated polemical goal is that they portray the claims of past philosophers through oversimplified, uninformed caricatures that already assume what embodied realism sets out to prove. Thus, they dismiss as mistaken countless legitimate achievements that could buttress their own position, were they not so intent on portraying second-generation cognitive science as the only correct philosophical method. (This "second generation" of cognitive scientists rejects "the fundamental tenets of traditional Anglo–American philosophy" (p. 75) as mistaken "on empirical grounds." The authors admit this empirical "evidence" is really a disagreement over basic—a priori!—"methodological assumptions" (p. 78). No wonder they often use words such as *argues* (p. 81) where they mean *assumes*.) A short review cannot mention all the instances of these errors in such a lengthy book—

it could have been considerably shorter than 600 pages had the authors not adopted an annoyingly repetitive style. Instead, I shall focus on the authors' treatment of Kant, whose *defense* of a priori knowledge they never actually consider.

That Kant's failure is crucial to the authors' success becomes apparent when they *declare* him mistaken in the opening pages (p. 5): "There exists no Kantian radically autonomous person and a transcendent reason that correctly dictates what is and isn't moral. Reason, arising from the body, doesn't transcend the body." They impute to Kant a theory of "absolute freedom" that Kant decries as a transcendent (*unknowable*) idea; moreover, they show no awareness of the fundamental thesis of Kant's first *Critique*, that "All our knowledge begins with experience." Whereas Kant infers a priori forms as necessarily *arising out of* our embodied experience, the authors somehow *know* that "There is no a priori" (p. 5)—a patent absurdity to anyone who "gets" Kant's a priori. The authors assume (a priori) that we have no a priori knowledge, the "validity" of philosophical theories being dependent on "empirical confirmation" (pp. 7, 256). They claim (my emphases) that "our conceptual systems draw *largely* upon the commonalities of our bodies" (p. 6), such that "abstract thought is *mostly* metaphorical" (p. 7) and "answers to philosophical questions have always been, and always will be, *mostly* metaphorical." Yet they provide no explanation for the "more" (the formal grounding of embodied knowledge) implied by these claims; their a priori rejection of the a priori prevents them from doing so.

Part I's refutation of "classical metaphysical realism" is correct, though naive from a Kantian perspective. In their search for "basic-level categories" (pp. 26ff), the authors conflate intuition (embodied input) with conception (mental processing of that input). Showing their ignorance of Kant (and others), they claim that in the "faculty psychology" assumed by "the Western philosophical tradition" (p. 36)—as if only one such tradition exists!—"no aspect of perception . . . is *part* of reason" (p. 37). But Kant's threefold synthesis, wherein all perception first occurs, *is* a rational process of understanding, not raw sensibility. By neglecting Kant's epistemology, the authors lose a golden opportunity to affirm how a major player in "the tradition" developed a system that dovetails nicely with cognitive science. Assuming such issues must "be settled in experimental neuroscience, not in the arena of philosophical argumentation" (p. 43), the authors never defend their empirical bias with reasons; demonizing opponents as "conservative", they base their rejections on wild generalizations (often without citing relevant literature), such as that in analytic philosophy "all concepts are literal and there are no such things as metaphorical concepts" (p. 87). Such ad hominem attacks make good rhetoric, but will not fool anyone who sees through the deceptions of self-referential arguments—the claim to have "no a priori commitments" (p. 80) itself being an a priori commitment. They repeatedly employ the same non sequitur: Because human knowledge consists of *some* A (e.g., metaphors), it therefore *cannot also* consist of some −A (e.g., non-metaphorical reality).

The authors rightly affirm the (Kantian) insight that the stability of science does not require believing "that science provides the ultimate means of understanding everything" (p. 89), but wrongly infer that this entails rejecting *all* a priori philosophizing. Their fundamental assertion, that "we never were separated or divorced from reality in the first place" (p. 93), does not preclude Kant's proof that the embodied mind employs a priori forms that can be regarded (through abstract reasoning) as distinct from the world "in the second place". Few have held the extreme positions the authors impute to past philosophers; most (like Kant) would *agree* with the authors' affirmation of "distinct 'truths' at different levels", with "no perspective that is neutral between these levels" (p. 105). Fortunately (but inconsistently?), they claim their brand of physicalism, with its tracing of all past philosophical theories to "folk theories" grounded in metaphor, is not "eliminative" (pp. 109–114). Unfortunately missing is a discussion of the *coherence* theory of truth as an alternative to the correspondence theory they refute (e.g., pp. 98–102, 165–166, 199); instead, they misleadingly call their quasi-Kantian position "an *embodied* correspondence theory of truth" (p. 233).

Chapter 14 demonstrates that much of our common moral reasoning is essentially metaphorical (e.g., relating the good to health, wealth, or family relations), asserting that the "structure and logic" of moral theories "come primarily from the source domains that ground the metaphors" (p. 329). The authors summarize the major moral philosophies in terms of how they use (or *could* use) a basic "family" metaphor that conveniently favors their own (liberal) "Nurturant Parent" position—a bias that could carry weight only if the metaphor is structured by some pure, a priori concept. Their structuring assumption is (paradoxically) that ascertaining the "universality" of such metaphors requires "cultural research" (pp. 311–312), there being "no pure moral concepts" (p. 328). Insisting the choice of moral principle "is an empirical issue" (p. 332), they confidently label a whole range of moral theories as simply "wrong" (p. 331). Must our moral reasoning *end* in metaphor simply because it begins there?

Chapter 20 explores how a network of moral metaphors undergirds Kant's moral philosophy. Here, as throughout Part III, the authors exhibit masterful eisegesis, employing what we might call (using their own tactic against them) an "Unconscious Assumption Is Making Use Of" metaphor to "demonstrate" that metaphors operate even when no textual evidence supports the claim. To justify their rejection of deontology, they (mostly) paraphrase selective passages that exhibit so-called "Strict Father Morality", entirely ignoring texts such as *Religion within the Bounds of Bare Reason* (1792), where Kant affirms the authors' preferred "Nurturant Parent Morality" (pp. 416–417). Anyone who has actually read Kant will find the "Kant" presented here—who allegedly grounds evil in the body (pp. 426–427, 433–434), for example—virtually unrecognizable. While the authors point out some interesting metaphors operating in Kant's texts, nobody should read this chapter for insight into what Kant *meant*; for this "Kant" is mostly a figment of the authors' fertile imaginations, a product of *their* folk theory of meaning, that Metaphors Cause Theories. Yet it is just as plausible to assume metaphors merely *correlate* with relevant theories. Apparently unaware that Kant's whole project (not unlike their own) was to *critique* the power of pure reason (pp. 539–540), the authors' argument boils down to a mere repetition of their folk theory that *empirical* evidence somehow disproves the a priori: "cognitive science ... invalidate[s] the central thrust of [Kant's] theory" (439) because "pure reason ... does not and cannot exist."

Many of this book's 25 chapters (especially Part II) provide illuminating summaries of recent research in cognitive science, naming numerous types of basic metaphorical relation; but to portray this web of empirical relations, the "cognitive science of philosophy" (p. 338), as *itself* a form of philosophy capable of "assess[ing] the adequacy of philosophical theories" (p. 340), is a major category mistake—as if naming a coherent system of metaphors magically constitutes access to ultimate reality! This leads to absurd, elitist claims, such as that we must "understand our unconscious moral systems" in order "to act morally" (p. 341)! What the authors *meant* to say, hopefully, is that acting morally requires *attunement* with this unconscious (noumenal?!) system—precisely Kant's position. They claim, inconsistently, that *because* "Kant's morality ... is irreducibly metaphoric", we must "abandon Kant's claim that morality issues from a transcendent" source (p. 345). Yet they themselves elsewhere affirm a form of agent-causation (p. 224) not unlike Kant's metaphor of "noumenal" causality. Is perspectival reasoning allowed only for cognitive scientists?

Those chapters focusing on past philosophers (mainly Part III) will not convince those well-versed in the positions being attacked. Unconsciously employing what I call the "Analyzing Metaphors Is Empirical Research" metaphor, the authors respond to opponents not with reasons, but with force (p. 71): "The answer is a loud 'No!'" The chief lesson to learn from this book's failure to supplant Western philosophy is that appeal to empirical facts (such as the admittedly metaphorical nature of most philosophical concepts) is no excuse for sloppy reasoning. Philosophy would not exist if all truth were *merely* as the authors say it is. In casting aside the *structuring* function of the a priori, the authors delude themselves, claiming to present a philosophy that is "close to the bone" (p. 8), when in fact what they provide is *only* a "philosophy in the flesh", sorely in need of a skeleton, such as that offered by Kant or the numerous other genuine philosophical systems they claim to refute.

STEPHEN R. PALMQUIST
Department of Religion and Philosophy
Hong Kong Baptist University
Kowloon, Hong Kong
stevepq@hkbu.edu.hk

Neither Brain nor Ghost: A Non-Dualist Alternative to the Mind–Brain Identity Theory by W. Teed Rockwell. MIT Press, A Bradford Book (Cambridge, MA), 2005 (hardcover), 2007 (paperback). 253 pp. $20.00 (paperback). ISBN 978-0-262-681674.

Radical Embodied Cognitive Science by Anthony Chemero. MIT Press, A Bradford Book (Cambridge, MA), 2009. 272 pp. $30.00 (hardcover), $22.95 (paperback). ISBN 9780262013222.

Eighty-five years ago in his major work *Experience and Nature*, the American philosopher John Dewey wrote the following:

> At every point and stage . . . a living organism and its life processes involve a world or nature temporally and spatially "external" to itself but "internal" to its functions. (Dewey, 1925:278)

This succinct idea carries within it the originating premise of the Hypothesis of Extended Cognition (HEC) and the Hypothesis of Radical Embodied Cognition (HREC), cutting-edge theories of cognitive science held by W. Teed Rockwell and Anthony Chemero, respectively. The argument is that the world outside the body does not consist merely of *objects of* cognition; instead, these "external" factors are *internal to the self*. As such, they are functional elements of any cognitive system and are indispensable to such systems. In other words, the dominant paradigm of the nature of mind and cognition is called into question.

HEC and HREC reject the mind–brain identity theory (MBI). Consciousness resides instead in a dynamic field that includes the brain, the body, and the world: "Even the most private, subjective, qualitative aspects of human experience are embodied in the brain–body–world nexus" (Rockwell, p. 158). Rockwell refers to this as a "behavioral field." He pulls no punches when it comes to following out this hypothesis to its inevitable conclusion:

> At any given time, there is a region within my world . . . within which everything is ready-to-hand for me. . . . And this region, I maintain, is *me* in the most unambiguous sense possible. (Rockwell, p. 106)

This *region* (which is on that theory a person) does not include the entire world. It has flexible boundaries depending on the range of activities and interests marking out the interrelations of an individual brain–body system with the world (Rockwell, p. 107). Since it is a dynamic field of processes, Rockwell and Chemero both believe it may be theoretically described by means of Dynamic Systems Theory (DST). Rockwell sketches the main outlines of DST, and Chemero's book goes into greater technical detail on this point.

As the entire discussion resides in a fomenting interface between philosophy and cognitive science, it is desirable to recall what issues, scientific and philosophical, lie in the background.

Critique of Presuppositions in Cognitive Science

Rockwell and Chemero argue that the field of cognitive science abounds in the design of experiments and interpretations of experimental results based on presuppositions that do not stand up under serious analysis. Rockwell presents multiple examples of these sorts of difficulties beginning with pointing out the impossibility of separating the cognitive role of the brain from that of the nervous system as a whole.

The theoretical separation of the brain from the nervous system assumes that the brain is a CPU-like controller, while the rest of the nervous system functions merely as sets of convenient cables carrying input to the brain. It is in the brain where actual cognitive functions take place, in the form of computational manipulation of representations of the outer world—what Chemero refers to as "mental gymnastics."

The brain-as-CPU theory is what would be left over after Descartes' idea of a disembodied soul mysteriously connected to the brain is disavowed. Rockwell refers to this as Cartesian materialism. Rockwell cites researcher Patricia Churchland as asserting unequivocally that the "mind is the brain" (P. S. Churchland, 1986:ix); then Rockwell shows that actual research results as well as interpretations by researchers do not support the separation of the brain from the rest of the nervous system with respect to cognitive functions (Rockwell, p. 23).

Rockwell's ultimate goal is to show that just as one cannot draw an absolute line between the brain and the rest of the nervous system, or between the brain-plus-nervous system and the body as a whole, so it is not possible to make a valid separation of the brain–body system from the world. In the course of his effort he constructs a variety of fictional scenarios illustrating some point or other about what we do or do not mean by key concepts in use in interpretive conclusions. Thus we find him talking at length about "zombies," "pink ice cubes," and "twin Earths," in scenarios that may strike impatient (non-philosophical) readers as absurd, tedious, or simply irrelevant because of references to impossible situations.

However, such philosophical excursions are sometimes necessary to make a point. Rockwell is struggling here against bias among mainstream cognitive scientists (as well as many others in other fields and even in the general public, where MBI is commonly accepted). So he takes the risk of seeming too esoteric and repetitive in favor of a very real need for persuasion in a matter that in essence demands a paradigm shift across the whole spectrum of our understanding of mind.

One more direct analysis that Rockwell provides is his chapter on "Causation and Embodiment." Here he undertakes a careful critique of views of causality in relation to cognitive science. Rockwell points out that "Because there are many crucial things happening in the brain every time we feel or think, neuroscience naturally assumes that brain activity is the sole cause of mentality." This assumption, he notes, is because "the goal of neuroscience is to discover the brain events that participate in the causal nexus responsible for mental events." The common assumption is that those brain events are the *sole cause* of their mental correlates—the conclusion being that the mind resides wholly in the brain (Rockwell, p. 54).

Rockwell maintains that the notions of *atomistic causality* and *intrinsic causal powers* both support the dominance of Cartesian materialism in cognitive science. Atomistic causality proposes that "a single event produces a causal relationship with another single event, and this connection could be completely independent from any other fact in the universe." On that view, single events in the brain would stand in a direct causal relation to some single mental states. Against this, Rockwell cites Mill as holding that "causes cannot be separated from their context of conditions."

Rockwell's point is well-taken. Interpretations of results in neuroscience research abound in examples of such separation. Any portion of the brain that is active during some process, such as remembering, is automatically identified by the neuroscientist as the sole cause. Typical interpretations are that the "memory" is "encoded" in the active region of the brain.

However, even if atomistic causality is eschewed, it could still be argued that the brain is a system that itself possesses intrinsic causal powers (Rockwell, pp. 55–57). Against this, Rockwell argues that causal properties are "fundamentally relations, not monadic predicates." Rockwell concludes that "the causal nexus that is responsible for the experiences of a conscious being is *not* contained entirely within the brain of that being" (Rockwell, p. 58, his italics). This sort of holistic position, repeated many times over with respect to different aspects of MBI theory, is well-represented and constitutes a substantial portion of his book.

In critiquing the presuppositions inherent in much of mainstream cognitive science, Anthony Chemero takes a more direct approach. He describes a flawed logic, which he refers to as "Hegelian argument" (after the philosopher G. W. F. Hegel, 1770–1831). Hegelian arguments begin with a set of premises based on some predetermined conceptual framework without empirical foundation, and then conclusions are drawn from those premises contrary to empirical evidence. Hegel's argument, for example, was to prove that "no eighth planet can be discovered." (Chemero might include among such arguments the one sometimes attributed (perhaps falsely) to Aristotle, to the effect that flies must have four legs because two is not enough, three is an imperfect number, and more than four are unnecessary.)

At any rate, Chemero points out that most philosophers and scientists are wary of such conceptual arguments, but that ". . . this attitude has not made its way into cognitive science, where conceptual arguments against empirical claims are very common." Chemero goes even further. He says "Indeed, one could argue that the field [of cognitive science] was founded on such an argument" (Chemero, pp. 4–6). In other words, he holds that there is something *fundamentally* rotten in the state of cognitive science.

Chemero's first illustration of this claim is to show how a key argument of Noam Chomsky's commits this logical blunder. Chomsky's argument is "the first in a string of Hegelian arguments in cognitive science." Among these is the "systematicity argument" set forth by Fodor and Pylyshyn (1988) which Chemero characterizes as "one of the most important and influential in the recent history of cognitive science" (Chemero, pp. 7–8). Chemero's discussion of such logical flaws, which seem to amount to a kind of intellectual disease, is all the more powerful because of the clarity of organization and expression with which he presents his analysis.

Next, Chemero strikes at the heart of the matter by explaining what makes the HREC alternative, in broad opposition to MBI, *radical*. HREC carries the tenet of HEC that cognition resides in dynamic brain–body–world systems a step further by denying that cognitive processes within such systems operate by manipulating *representations* of the world. Thus to the degree that cognitive science is "representationalist" (and hence computational) it is on the wrong track. Formally, HREC is expressed as the claim that the tools for explaining embodied cognition, which include DST, do not require the positing of mental representations.

Chemero represents this view as descending from American naturalism as exemplified by Dewey and James, and from Gibsonian ecological psychology. These he says are *eliminativist* views (i.e. inner representations are eliminated from theory of cognition) (Chemero, pp. 29–30). Although HEC also invokes the brain–body–world nexus, HEC still may embrace some form of representationalism, and would thereby be committed to a form of computationalism in that cognitive activity is seen as manipulation of representations, but occurring somehow within a brain–body–world nexus rather than in the brain alone.

Philosophical Considerations

Rockwell's book has as its subtitle "A Nondualist Alternative to the Mind–Brain Identity Theory." Rockwell rejects Cartesian dualism, but he also rejects Cartesian materialism, particularly its extreme form, "eliminative materialism." This is the view that all so-called mental activity is nothing more than physical states of the brain, and therefore "there are no such entities as thoughts or sensations, and never were." Those who think they possess a mind, or are a conscious thinking self, are suffering from a kind of illusory "folk psychology"(Rockwell, p. 5).[1]

One might wonder what form of madness could bring such seemingly adept thinkers as Paul Churchland (1989), as cited by Rockwell, to adopt such a view. In effect, anyone who espouses eliminative materialism is *denying his own denial*. It is, in fact, a separation of oneself from oneself, and is one of the adverse results of the dualistic view. Once a single absolute separation of one element of experience from another is accepted, the world *shatters* to the extent that a person cannot understand either who, or where, or what he or she is. This is a psychological concern and is not limited to theoretical issues in cognitive science. Ecological psychologist Robert Greenway, for example, refers to dualism as the "radical wound" caused by "distinctions that become disjunctions" (Greenway, 2009). If this dimension of the discussion is taken into consideration, to the degree to which cognitive science abets and perpetuates a dualistic separation of mind from world, it is on negative *ethical* ground.

From this standpoint, it is unfortunate that Rockwell's book makes no mention of this broader socio–psychological context. There is a suggestion of this context in Chemero's final chapter on the metaphysics of HREC, but it remains within an abstract philosophical discussion rather than that of an existential malaise. The fundamental reason for denying representationalism should be, in this present writer's opinion, that once the psyche of an individual personality is collapsed into the sphere of computational manipulation of pictures of the world, there is no longer a lived world, and the relation between an individual and his or her milieu can become destructive—so much so, in fact, that we have the absurdity of the eliminative materialist denying his own denial even as he is denying it.

Here we crash head-on into a problem with both HEC and HREC. The following questions raise their unpleasant countenances: First, whether "having" representations is the same thing as having a mind. Second, whether eliminating representations amounts to the eliminative materialist view that our sense of having a mental life of any sort at all is simply an illusion. Third, whether it makes any sense at all to talk about "representations" being "embodied" by either a brain *or* a brain–body–world dynamic system (choose whichever you prefer).

The competition between representationalist and nonrepresentationalist views of cognition is a rather loud echo of the kinds of philosophical clashes found throughout the history of both philosophy and science, such as that between nominalism and realism, rationalism and empiricism, or indeed the clash of form and matter. Is mind reducible to physical states of the brain, or of the brain plus body, or of a brain–body–world dynamic system, such that within any such system there exist, or do not exist, substructures that somehow "represent" other structures "outside" the system? And can any such physical system *be a mind* and not contain within itself any representational structures at all? This opens a philosophical/scientific Pandora's box, which both MBI theorists and HEC/HREC advocates are going to have a great deal of trouble shutting, if that is even possible. (At the risk of mixing metaphors, I'm tempted to start talking here about the Sorcerer's Apprentice, but let be).

The undermining presence of issues like those above are revealed at once. In Rockwell's discussion we have a confusion of terminology about what it means to say a *mind* is *embodied* in a physical system. There are two difficulties here: The first is how "mind" is adumbrated, and the second is how "embodied" is understood. On just two pages, Rockwell characterizes mind as "mental properties," "mentality," "a mental system," and "mental processes." So is mind a quality (mentality), a process, a property, or a system, or all of them at once? But as things go on, the favored characterization of mind is that it is constituted by a *set of properties* which might also be termed mental states.

Thus Rockwell refers to "our visual mental states," by which he seems to mean something like "having a visual perception" or perhaps just "seeing something." Included also are thoughts and beliefs, "knowing or remembering a fact about the world," and so on, all of which either constitute the mind or are properties of a mind, or are properties of either the brain or a dynamic brain–body–world system, such that we say one or the other of these is, or has, a mind. (Rockwell, p. 71)

Now there is a reason that Rockwell shifts from (infrequently) talking about a mind to (frequently) talking about sets of properties. This is because of his concept of what it means to say that mind is embodied. He seeks to resolve the question of dualism by determining how we can contain the concepts of mind and body in a single system. This must perforce be a physical system "embodying" a mind, the latter seen as a set of properties. But these cannot be physical properties, else we are committed to eliminative materialism. They remain mental properties—sets of properties of a peculiar sort called "mental."

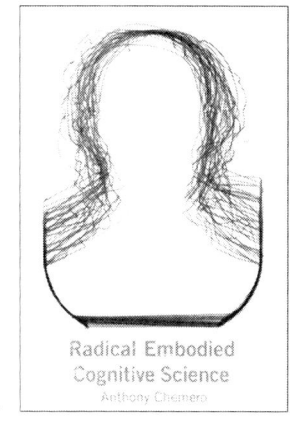

The underlying reason for this sticky philosophical position, which avoids the question of how any set of properties constitutes the mind of a conscious self, is that Rockwell believes the problem is solved by invoking the concept of "supervenience," which is admittedly a "scrupulously downsized technical term with some similarities to both causation and identity." (Rockwell, p. 69).

Thus if it is understood that the mind *supervenes on* the body, according to Rockwell the problem of dualism is resolved. And supervenience, in turn, has to do with sets of properties. So the convenience of supervenience is to view the mind as a set of mental properties, and then to assert that these supervene on a set of physical properties. By definition, this entails that any two individuals who are indistinguishable in their physical properties must perforce exhibit identical mental properties, and if two individuals possess different mental properties, they must also have different physical properties. But the relation is not symmetric: Two individuals may have the same mental properties but differ in their physical properties.

The relation of supervenience may cogently be asserted in certain kinds of physical situations. For example, it may hold among relations of the acceleration, velocity, and position of an object in space. But there are two problems with the assertion of supervenience as the relation between sets of mental properties and sets of physical properties. The first is that it is empirically impossible to determine whether it is true that "any two individuals indistinguishable in their physical properties have identical mental properties," because there *are* no two individuals who are indistinguishable in their physical properties—their fingerprints alone will testify to this (not to speak of DNA).

What Rockwell really has here is a premise, namely the premise that the relation of mind (as mental properties) to body (as physical properties) is that of supervenience. And that premise is purely conceptual with no empirical foundation. Any arguments Rockwell presents as following from such a premise are instances of Chemero's Hegelian arguments. Bluntly speaking, the invocation of supervenience as a solution to the mind–body problem is nothing more than a piece of philosophical thaumaturgy.

The second problem, which really brings us to the nub of the matter, is the use of the term *individuals* in the definition of supervenience. The utterly devastating difficulty of the entire enterprise has to do with the way it systematically uses, but systematically discounts, the nature of the self. For how do we determine if two individuals have, or do not have, the same properties, unless we presuppose that they are indeed individuals? And presumably to be individuals, it must be predetermined that they have minds.

So, for example, if we say that "the mind is a set of mental properties," we have committed a logical mistake, because to know that a property is mental we must already know that it is a mind that has those properties. We have already encountered this difficulty in Rockwell's statement of what it is that he thinks he is. When he says "there is a region within *my* world in which *I* am engaged ... and this region, *I* maintain, is *me*," he presupposes that he is a *self* who can formulate hypotheses as to what he is (my italics). It is significant that Rockwell very seldom mentions "self" in his discussions but prefers to focus on mental properties instead.

A related case appears in Chemero's discussion of direct perception. He introduces the "problem of two minds" which arises when perceptions are direct, rather than mediated by internal representations. The problem is that if two separate individuals A and B *directly* perceive the same object X, the object will be part of both A's perception and B's perception, and thus at odds with the assumption that minds are private. The peculiarity arises when Chemero *agrees* that the minds involved must be private, and argues that since perception is a relation, and since the relations of A to X and B to X are different, so their perceptions "do not overlap." Chemero is therefore unwilling to give up the alleged privacy of minds—but at the cost of now having two X's, the X that A perceives and the X that B perceives.

The problem here is the reverse of that in the Rockwell case. Instead of not taking into account the role of self in defining mind and consciousness, Chemero posits that selves are "private" and hence cannot share an object of perception with any other individual, on the ground that any perception is a unique relation between an observer and a thing observed. But this is to render meaningless the entire enterprise of extending, or embodying, the mind into a system of interactions between brain-body and world, because the things in the world that are part of any such system are necessarily shared by multiple overlapping systems. Indeed, the very possibility of such human experiences as those of empathy, friendship, knowledge of others, and love, depend on some sort of shared, not private, self. Neither the privatized self implied by Chemero, nor the self-as-bag-of-properties advocated by Rockwell, can satisfy this profoundly necessary requirement.

The upshot of this discussion is that the less philosophy of mind occupies itself with the technological details requisite for creating a simulated intelligence, and the more it remains sensitive to the full spectrum of human experience, the better. The advocated use of DST in cognitive science may indeed have useful results of a certain limited sort; but it will almost certainly not solve the problems that center about the nature of the self and the presence of dualism in culture and psychology.

STAN V. MCDANIEL
Professor of Philosophy Emeritus,
Sonoma State University
Rohnert Park, California, USA
stanmcd2@sbcglobal.net

Notes

[1] The question arises as to why this reduction of "mind" to nothing but physical substance does not qualify as *nondualistic*; actually, eliminative materialism is not a genuine nondualist philosophy, because instead of reconciling properties of mind with those of matter, it simply denies the existence of mind in any recognizably ordinary sense. The aim of a nondualistic view is to *reconcile* consciousness with matter.

References

Churchland, P. M. (1989). *A Neurocomputational Perspective.* Cambridge, MA: The MIT Press.
Churchland, P. S. (1986). *Neurophilosophy.* Cambridge, MA: The MIT Press.
Dewey, John (1925). *Experience and Nature.* New York: Dover Publications Inc. (reprint, 1958)
Fodor, J., & Pylyshyn, Z. (1988). Connectionism and the cognitive architecture: A critical analysis. *Cognition, 28,* 3–71.
Greenway, Robert (2009). Interview in *Ecopsychology*, March 2009, p. 47.

Synchronicity: Multiple Perspectives on Meaningful Coincidence edited by Lance Storm. Pari Publishing (Pari, Italy), 2008. 358 pp. $18.95 (paperback). ISBN 978-88-95604-02-2.

Is synchronicity exclusively an acausal event? Is it similar to psi? Can it be subject to experimentation? Is its phenomena best understood through the lens of quantum mechanics, as a product of the *Unus Mundus*, or as an emergent function of the Self on its compensatory drive toward individuation? Do divinatory practices such as the *I Ching* offer the potential to empirically capture its sporadic occurrence?

These are just a few of the questions raised in this collection of some of the more outstanding papers presented on synchronicity over the past 40 years. Compiled and edited by Lance Storm, they're organized into six sections, with each addressing synchronicity from a particular vantage point.

Storm states at the outset that he wants to revisit synchronicity, not "from a flight of fancy, but from the well-considered, well-developed ideas that spring forth from the fountainhead of those whose work comprises this anthology" (p. xv). In the Foreword, Robert Aziz likens the book to attending a conference on synchronicity consisting of presenters with "highly specialized and diverse professional backgrounds" (p. xix). I believe with this compilation of papers, Storm achieved his goal, and, fitting with the Aziz analogy, after reading them I felt that post-conference inspiration to integrate the many specialized viewpoints.

In the first section of the book, "The History and Philosophy of Synchronicity", Kenower Bash's Chapter 1, "The Improbable Jung", provides an overview of Jung's discoveries. Bash takes us back to Jung's brilliant essay as a schoolboy, his later research on word association, psychological types, and internal psychic processes. Throughout the paper, he shows how Jung's nature, experiences, and courage contributed to his discoveries of synchronicity.

In Chapters 2 and 4, David Peat and Marialuisa Donati examine the specific relationship between the Nobel physicist Wolfgang Pauli and Carl Jung. Peat discusses how Pauli's influence on Jung and additional contact with Albert Einstein provided him with the foundational scientific support for his evolving theory on synchronicity. Donati also addresses the collaboration between Pauli and Jung from a more personal perspective. She notes how Jung assisted Pauli to identify his feelings and dream images and integrate them with his very one-sided intuitive-thinking style. Conversely, she relates how Pauli helped Jung focus his attention on the ontological and archetypal character of synchronicity, going beyond its previously phenomenological and empirical emphasis.

Roderick Main's paper, "Religion, Science, and Synchronicity" (Chapter 3), explores the role synchronicity played in Jung's understanding of the relationship between science and religion. He examines Jung's clergical family influences, his fascination with J. B. Rhine's ESP experiments, and how Chinese philosophy and the *I Ching* all influenced his conception of synchronicity. Main contends that Jung sought to restore meaning in science and to bring empiricism to religion.

The second part of this anthology is labeled "Synchronicity in Practice". In Chapter 5, "Synchronicity and Telepathy", Berthold Schwartz shares some anecdotal synchronistic events arising from a clinical context. He discusses what appear to be telepathic components in those events, and how synchronicities in a clinical setting may be of therapeutic value. In the next chapter, "Synchronicity, Science and the I Ching", Shantena Sabbadini describes the relationship between synchronicity and the ancient divinatory process of the *I Ching*. She notes Jung described its contents as a "catalogue of sixty-four basic archetypal patterns" (p. 80), and viewed its process as synchronistic. Like Main, she also elaborates on the Chinese philosophy so critical to understanding the *I Ching* and synchronicity.

Last in this section is William Braud's Chapter 7, "Toward a Quantitative Assessment of 'Meaningful Coincidences'". There he describes two experiments that produced significant results for evidence of synchronicity. Despite his results, however, he questions if they might be explained by conventional forms of psi, suggesting like Schwartz a possible relationship between psi phenomena and synchronicity.

Following the section on practice, the third part of the book called "The Ontology of Synchronicity" consists of two significant papers by John Beloff and Charles Tart. In the eighth chapter, "Psi Phenomena: Causal versus Acausal Interpretation", Beloff distills synchronicity down to what he considers its critical ingredients. For instance, he states: "So far as Jung himself was concerned it was not just any old coincidence, however startling, that exemplified the principal of synchronicity, it was specifically those conjunctions which exhibited archetypal ideas" (p. 101).

In contradiction to Jung, Beloff resists the connection Jung made to J. B. Rhine's ESP research where psi and synchronicity are equated, with both considered acausal. Beloff's position is that a causal nexus must exist in synchronicities even if that cause may still remain hidden. That type of cause, he concludes, may resemble something more like Aristotle's formation or final cause.

In Chapter 9, "Causality and Synchronicity: Steps toward Clarification",

Charles Tart continues to examine the concept of causality. Beginning with perceptions in infancy, he distinguishes the initial temporal and proximal perceptions of causes in the physical world from causes determined psychologically. Tart lists and elucidates eight types of causality and two types of pseudo-causality. Tart further cautions that should experiments prove difficult to replicate when conducting psi research, we not give up seeking causes for evident psi manifestation. He believes it's "intellectually lazy" to relegate them to the acausal workings of synchronicity.

The fourth part of the book is titled "The Synchronicity Debate". In Chapter 10, Mansfield et al. analyze previously unavailable letters between Carl Jung and J. B. Rhine and examine the relationship between psi phenomena and synchronicity. They grant that in some cases, similar to Schwartz's telepathically facilitated anecdotes, psi phenomena may accompany synchronistic events, but they firmly believe the two phenomena are distinct. Mansfield states:

> Parapsychological experiments are acausal in the Jungian (historical causality) sense, but exhibit "scientific causality", since repeatable connections between mental and physical events can be reliably established.... On the other hand, since synchronicity is a sporadic, nonrepeatable expression of the Unus Mundus, the unitary ground underlying both matter and psyche, it is both historically acausal and scientifically acausal." (p. 142)

Mansfield underscores what he calls the "*sin qua non* of a synchronistic experience" and that which further defines the difference between synchronicity and psi: its "archetypal meaning". This type of meaning, he suggests, is very different from the meaningful correlation of a statistically significant psi result. He speculates that possibly due to Jung's initial trepidation of coming out to the scientific community with a controversial idea such as synchronicity, he may have conflated synchronicity with the emerging positive patina of Rhine's parapsychological research.

Chapter 11 is Lance Storm's first paper in the book. Appropriately following the Mansfield et al. paper, it was originally written as a response to their position. It is entitled "Synchronicity, Causality, and Acausality". Storm first sets out to clarify Jung's notions of synchronicity, re-contextualizing some previously extracted and distorted attributions made to Jung. He then addresses a few other arguments that are typically employed to discredit the legitimacy of psi, and of synchronicity. Among these are the Law of Large Numbers, retrieval cues, and the philosophic semantic jockeying over words such as *meaning*, *cause*, and *contingence*.

Storm's central position differs from Mansfield's. He sees parallels between synchronicity and psi, and outlines phenomenological similarities between the two. For example, he draws a parallel between the archetypal contingencies of

Book Reviews 339

a synchronistic event and the psi-permissive and psi-conducive contingencies of psi phenomena. He also points out that initially both psi and synchronicity are chance-like (acausal) in nature. And yet, because Jung and Braud succeeded in experimentally researching synchronicity, he believes psi and synchronicity share scientific causality. Lastly, Storm believes that beyond the statistical meaningfulness of psi research, psi phenomena can also produce meaningfulness similar to synchronicity, and it too can lead to personal transformation.

John Palmer continues examining the relationship between psi and synchronicity in Chapter 12. In his paper, "Synchronicity and Psi: How Are They Related?", he initially addresses the key ingredients that Jung states make up a synchronistic event: time and meaningfulness. With regard to time, he notes how Jung modified his own original position for the requirement of simultaneity of the synchronistic events, later allowing for a synchronistic event to be perceived prior to its occurrence. Palmer believes this shift opened the door for psi experiences such as precognition and telepathy to be included in synchronicities.

On the issue of meaningfulness, Palmer feels that it is nonsensical to say that meaningfulness is intrinsic in the corresponding events. He believes the meaning is subjectively projected by the individual and individual subjectivity can be diminished if the meaning is arrived upon by a consensus.

Palmer disagrees with Mansfield's suggestion that the volition employed in psi research further separates synchronicity from psi. In that regard he suggests not conceiving volition as causing synchronicity, but as accompanying it. Palmer also discusses two theories of psi called the transmission and correspondence models. He sees synchronicity most closely fitting the latter, particularly the conformance model of Rex Stanford. Finally, Palmer leaves open the possibility that with the proper perspective and controls, synchronicity like psi may be able to be empirically researched.

The next three chapters compose the fifth part of the book called, "New Conceptions of Synchronicity". In Chapter 13, Lila Gatlin presents "Meaningful Information Creation: An Alternative Interpretation of the Psi Phenomena". Her basic position is that there is no transmission of meaningful information in a synchronistic event. There is no carrier, or signal. She believes the information involved has not been transmitted, but rather created via the mechanism of synchronicity, and that this process has had and continues to have evolutionary implications for our survival.

Chapter 14 is written by Joseph Cambray and entitled "Synchronicity and Emergence". Cambray draws from Ilya Prigogine's work on the study of "complexity". A Nobel laureate in chemistry, Prigogine articulates how order can emerge out of chaos, and in a similar fashion Cambray believes the meaning in a synchronistic event is emergent. He further states: "Synchronicities can be explored as a form of emergence of the Self" (p. 219). He also suggests that synchronicities offer "valuable clues" to the individuating psyche, but "they must be treated as value-neutral—that is, they do not in and of themselves convey direction to consciousness. Such direction can only come from reflective, ethical struggles with the meanings" (p. 229).

George Hogenson provides the last chapter in this section, "The Self, the Symbolic and Synchronicity: Virtual Realities and the Emergence of the Psyche". While also referencing Prigogine's notion of a "self-organizing concept in nature", and influenced by Cambray's articulation of emergence, Hogenson attempts to unify Jung's concepts of The Complex, The Archetype, and The Theory of Synchronicity under a single dynamic principal. He discusses power laws, phase transitions, and the "symbolic" as "more than a system of representations, but rather a relatively autonomous self-organizing domain in its own right" (p. 242).

The sixth and last part of this anthology called "Summing Up Synchronicity" consists of three chapters and an Appendix, all provided by Lance Storm. Here he attempts to clarify and integrate much of the work that precedes him in the book. In Chapter 16," Synchronicity, Science, and Religion", he first picks apart the definition of synchronicity and examines ways where meaning may be created, misread, or mistaken. Storm reminds us that Pauli and Jung "arduously and ardently maintained a connection between science and religion" and that in his opinion their unification would come by way of integration with the psyche as the underlying factor common to both (p. 256).

In the next chapter (Chapter 17), Storm focuses on "Archetypes, Causality, and Meaning". There again, he delves into the key components that compose and give rise to a synchronistic event. For example, regarding the argument of causal or acausal attributions made about synchronicity, he suggests employing the term *metacausal*, a blanket term for contingence, equivalence, and meaning, and one that also allows for an inconstant connection.

Finally in Chapter 18, "Synchronicity and Psi", Storm elucidates the many ways these two phenomena are similar. He takes a firm position that like psi, synchronicity may be able to be empirically tested, and he even suggests in the following Appendix some potentials the *I Ching* may offer in that regard.

I agree with Storm on the potential for synchronicity to be empirically investigated. Along with him, Jung, Braud, and Palmer, I believe with proper controls its phenomena may be captured and better understood. I don't agree with Beloff that attempting to do so would be a "fool's errand", or that it is "doomed to failure" as suggested by Sabbadini. Instead, I'm reminded that though Mansfield et al. initially doubted that synchronicity might reveal itself empirically, they ultimately backed off that position, suggesting that it may be possible if done "with the proper conditions and with sufficient rigor" (p. 150).

But conducting such empirical research would not only require rigor and proper conditions, it would also require remaining true to Jung's basic definition of synchronicity. Critical ingredients such as archetypal content would need to be activated and employed. For instance, as suggested by many, a certain amount of emotional intensity would need to exist in order to catalyze the process and constellate an archetype. This may be achievable in a clinical context. Another critical ingredient such as meaningfulness would need to be controlled. Following Storm and others before him, this may be accomplished by using the *I Ching* and its sixty-four hexagrams. Jung considered the *I Ching*'s hexagrams loaded with "archetypal meaningfulness".

Ingredients found in parapsychological research might also be utilized in an empirical investigation of synchronicity. Hypnosis and other psi-conducive and psi-permissive techniques might prove helpful. Employing the *I Ching*'s process of throwing coins, for instance, allows for randomness.

Storm states that "synchronicity and psi are the same fundamental process" (p. 295). It appears, therefore, that synchronistic phenomena like psi can benefit from being empirically investigated. Should such efforts bear fruit with synchronicity, a better understanding of its phenomena may not only evidence an interconnectedness of intrapsychic conditions, psi processes, and numinous assistance, but also contribute to our psychological, emotional, and spiritual growth.

Such a goal may seem lofty, but if we are mindful of Tart's admonitions that we not let "intellectual laziness" lead us to give up finding causes, who is to say what we may yet discover. If your interests are simply to gain a fuller understanding of synchronicity, however, I could think of no better book to provide the depth and breadth of perspective found in this anthology.

FRANK PASCIUTI
Licensed Clinical Psychologist
Charlottesville, Virginia 22902
frankpasciuti@hotmail.com

Laboratories of Faith: Mesmerism, Spiritism, and Occultism in Modern France by John Warne Monroe. Cornell University Press (Ithaca/London), 2008. 235 pp. $35 (hardcover). ISBN 9780801445620.

Writing in a scholarly and elegant style, this author presents part of the history of French heterodoxy: movements between science and belief of Mesmerism, Spiritism, Occultism, and Psychical Research in France between 1853 and 1925. The author is now Associate Professor at Iowa State University, and this book is based on his thesis at Yale University in 2002 (for which he was awarded Yale University's Theron Rockwell Field Prize).

The book title refers to the development and maintenance of spiritual beliefs through observation or experience of tangible facts. This link between subjectivity and objectivity is valid for both orthodox beliefs (e.g., miracles and the Christian's extraordinary; Sbalchiero, 2002) and new marginal beliefs. Nineteenth-century France is itself a valuable observatory of the opposition between Enlightenment and counter-Enlightenment movements that united in claiming a scientific ground. Advances in scientific knowledge rendered obsolete metaphysical concepts, such as the immortality of the soul, that could not be tested in the laboratory.

> For these believers, the only way to guarantee the continued validity of religion in the modern world was to radically change the texture of the human experience of the sacred. "Unique realism", in this view, had to come from a religious system that made the contemplation of the beyond into a scientific, empirical project. (p. 8)

French religious life has experienced a *crisis of factuality* which was solved by evolution in empirical evidence for some metaphysical propositions whose ramifications are still visible today (p. 3).

The book comprises five chronological chapters and an epilogue. Chapter 1 describes the 1853 arrival of American Modern Spiritualism in France. Many French had been intensely but for a short period fascinated by *turning* and *speaking tables*.

Chapter 2 returns to Mesmer's revolutionary therapy and occult science of animal magnetism, whose wonders and controversies since the late eighteenth century prepared the way for the reception given to Spiritualism. Monroe shows clearly how the Mesmerist movement has gradually been swept away by the vogue for Spiritualism, in part because the phenomena claimed by the Spiritualists were more stunning and persuasive that the magnetic phenomena. D. D. Home and other mediums from America acted as direct sources for the sacred, as missionaries for the new cause, and as scientific instruments of faith.

Chapter 3 shows the emergence in the mid-1860s and the rapid domination of the Spiritualist movement around its codificator Allan Kardec (Allan Kardec is the pseudonym for Hippolyte Léon Denizard Rivail). Kardec is a central figure in this book because he not only shaped the beliefs of his time but also the practices. His approach combining rigorous positivism and moderate anti-Catholicism is a powerful vehicle for a new vision of religious life, with both moral and political implications. The impressive sales of his writings and those of some of his disciples (Flammarion, Leymarie, Denis, and Delanne) made of Spiritualism the most broadcast popular philosophy in France at that time.

Chapter 4 describes the decline of Spiritualism following the death of Kardec, and a highly publicized trial that took place in 1875, in which several Spiritists were convicted of producing and marketing false spirit photographs. The following trust crisis in this kind of transcendental evidence led to a repudiation (for example, those claims of the astronomer Camille Flammarion) of the Spiritist doctrine as a solution to the crisis of factuality.

In Chapter 5 and the Epilogue, Monroe explores other heterodox movements: Psychical Research and Occultism. In their own way, these movements have proposed innovations in terms of beliefs, practices, and empirical evidence. The Epilogue is a brief overview of the years following the First World War and the emergence of a new heterodoxy: the founding of the Institut Métapsychique International in 1919 by a group of researchers between Spiritism and psychical research; the reprise by the surrealism of André Breton; the anti-heterodoxy of René Guénon; the fantastic realism around the book *The Morning of the Magicians* (Pauwels & Bergier, 1963); and the use by "cult milieu" of heterodox topics, such as with the Raelian cult and the belief in extraterrestrials.

Monroe approaches his subject with an exemplary academic neutrality when not mocking "evidence of things not seen." Believers in the paranormal are not shown as mere irrational people, and instead Monroe emphasizes the permanent rationality of their approach, as with Kardec. In this sense, this book illustrates a comprehensive sociology of the apparently most extreme beliefs, placing them in their political, cultural, and intellectual context. It does not reduce the complexity of these "scientifically unacceptable beliefs" (Irwin, 2009) to a deviation of thought developed by marginals, madmen, or idiots.

However, this neutrality will be frustrating for a scientific reader. Spirit seances in small groups, fraudulent photographs of the dead, and ingenious experiments by psychical researchers are put on the same footing: They are all "laboratories of faith." The "history of religion" perspective takes precedence over the "history of science" one. Thus, in this view, assertions such as "What made Palladino's phenomena remarkable was their audience" (p. 209) can divide. Gauld (1968) had already established that the Society for Psychical Research was founded by scholars during their "crisis of factuality". But this had not prevented him from showing the importance of empirical and theoretical progress engendered by the movement of psychical research. In Monroe's book, the treatment of this movement is rather thin: For example, Monroe says nothing about the Societé de Psychologie Physiologique, one of the first institutes for French academic psychology, which included the study of hypnotic and psychic phenomena. Nothing either on the Académie des Sciences Psychiques, founded in 1898 by theologians and scholars, and its *Revue du Monde Invisible,* which may have been relevant as a "great battle in which the Church can counteract the influence of scientism" (Guillemain, 2006:130). Monroe will even find "far-fetched" (p. 217) F. W. H. Myers' theories on the Subliminal Self from the point of view of a contemporary researcher. This retrospective judgment differs from the neutrality of the historian by extending controversies he is supposed to describe (for a contrary and complete view on Myers' theories, cf. Kelly et al., 2007).

Despite these minor reservations, this book gains a worthy place among a series of recent academic works dealing with various heterodox movements in France during this period (Edelman, 1995, Méheust, 1999, Le Maléfan, 1999, Plas, 2000, Lachapelle, 2002, Brower, 2005, Harvey, 2005, Sharp, 2006).

Renaud Evrard
Psychologist
Center for Information, Research, & Counseling on Exceptional Experiences
www.circee.org

References

Brower, M. B. (2005). *The Fantasms of Science: Psychical Research in the French Third Republic, 1880–1935.* [Ph.D dissertation, Rutgers University]

Edelman, N. (1995). *Voyantes, Guérisseuses et Visionnaires en France, 1785–1914.* Paris: Albin Michel.

Gauld, A. (1968). *The Founders of Psychical Research.* New York: Schocken Books.

Guillemain, H. (2006). *Diriger les Consciences, Guérir les Âmes. Une Histoire Comparée des Pratiques Thérapeutiques et Religieuses (1830–1939).* Paris: La Découverte.

Harvey, D. A. (2005). *Beyond Enlightenment: Occultism and Politics in Modern France.* DeKalb: Northern Illinois University Press.

Irwin, H. J. (2009). *The Psychology of Paranormal Belief: A Researcher's Handbook.* Herfordshire: University of Hertfordshire Press.

Kelly, E. F., Kelly, E. W., Crabtree, A., Gauld, A., Grosso, M., & Greyson, B. (2007). *Irreducible Mind: Toward a Psychology for the 21st Century.* Lanham, MD: Rowman & Littlefield.

Lachapelle, S. (2002). *A World Outside Science: French Attitudes toward Mediumistic Phenomena, 1853–1931.* [Ph.D. dissertation]. University of Notre Dame.

Le Maléfan, P. (1999). *Folie et Spiritisme Histoire du Discours Psychopathologique sur la Pratique du Spiritisme, ses abords et ses Avatars. 1850–1950.* Paris: L'Harmattan.

Méheust, B. (1999). *Somnambulisme et Médiumnité* (2 volumes). Paris: Les Empêcheurs de Penser en Rond.

Pauwels, L., & Bergier, J. (1963). *The Morning of the Magicians.* New York: Stein and Day.

Plas, R. (2000). *Naissance d'une Science Humaine, la Psychologie: Les Psychologues et le "Merveilleux Psychique".* Rennes: PUR.

Sbalchiero, P. (Ed.). (2002). *Dictionnaire des Miracles et de l'Extraordinaire Chrétien.* Paris: Fayard.

Sharp, L. L. (2006). *Secular Spirituality: Reincarnation and Spiritism in Nineteenth-Century France.* Lanham, MD: Lexington Books.

The Spiritual Anatomy of Emotion: How Feelings Link the Brain, the Body, and the Sixth Sense by Michael A. Jawer with Marc S. Micozzi. Park Street Press (Rochester, VT), 2009. 576 pp. $24.95 (paperback). ISBN 9781594772887.

The interaction between body and brain as it relates to emotion has long been an issue of debate within psychology, with the most recognized historical example in the field perhaps being the James–Cannon debate. Based on the premise that emotions are often accompanied by one or more somatic responses, the prominent psychologist William James (1884) had proposed that an emotion-arousing stimulus triggers a somatic response via the autonomic nervous system (ANS) that is then perceived and interpreted by the brain as a certain emotional sensation. From this perspective, emotion seems to have its prime source in the body. A similar theory was independently proposed by Danish researcher Carl Lange, forming the basis for the James–Lange theory. Physiologist Walter Cannon (1927), along with his colleague Philip Bard, later questioned the James–Lange theory and offered an alternative in which the stimulus triggers neural impulses that are sent to the thalamus, which are then routed both to the cortex and to the hypothalamus to initiate a somatic response via the ANS. In parallel with the activity of the ANS, signals from the hypothalamus travel to the cortex to initiate the emotional experience, which would then be accompanied by the associated somatic response. Thus, according to the Cannon–Bard theory, emotion seems to have its prime source in the brain, with the somatic response being a side effect. It would seem that, over the years, the emerging roles of the limbic structures in the medial temporal lobe have placed additional emphasis on the brain as being the likely source of emotion, with the contributions of the body receiving lesser attention. (On a brief personal note, this reviewer might add that this was the general perspective that he was taught when he first became a student of neuropsychology.)

In *The Spiritual Anatomy of Emotion*, emotion researcher Michael Jawer and physiologist Marc Micozzi attempt to reassess the interaction between body and brain in relation to emotion and shed light once again on the body's contribution. Through a review of the latest empirical and anecdotal evidence from fields such as medicine, biochemistry, neurology, neuroimaging, and psychoneuroimmunology in the opening chapters of the book, Jawer and Micozzi aim in part to make a case for consideration that, while the brain plays a major role in emotion, the body also has an important, but possibly overlooked, role. Jawer and Micozzi seem to offer a perspective similar to the one offered by molecular biologist Candace Pert (1997) in her book *Molecules of Emotion*. In Pert's view, the experience of emotion is guided by the biochemical actions of neuropeptides binding to various receptor cells found throughout the body that are thought to comprise a vast psychosomatic network, with a large amount of receptors being found in the limbic regions. Through this body-spanning network, mind and body are posited to work as one in experiencing sensation and emotion. Through their review, Jawer and Micozzi similarly argue that brain and body operate in tandem through various neurophysiological processes to give rise to emotion. (Like Pert, they use the term *bodymind* to refer to the proposed reciprocal interaction between brain and body.)

Relevant to the scope of this *Journal* is the way in which Jawer and Micozzi attempt to use this perspective to account for various forms of ostensible psychic (psi) phenomena. On the basis of a review of findings relating to the biological effects of electromagnetism, temporal lobe epilepsy, and the neurobiology of fear, the authors suggest that certain individuals ". . . may effectively displace the energy of repressed feeling into the surroundings" (p. 197), and that this may offer a possible explanation for reported apparitional experiences, haunts, and recurrent spontaneous psychokinesis (RSPK, or "poltergeist").

With respect to apparitions, the authors suggest that the energy of repressed feeling retained in a person's body, along with unresolved issues or preoccupations held within his or her brain, give rise to an apparition of the person who persists after death, seemingly consumed with these unresolved issues or preoccupations as their apparent purpose for remaining (p. 155). A potential weakness to this idea is that the parapsychological and psychical research literature on apparitions actually reveals very few cases that clearly suggest signs of conscious, purposeful intention on the part of apparitions (Roll, 1982, Sect. 2.4). Eleanor Sidgwick (1885) raised this point most succinctly in her review of apparition cases for the Society for Psychical Research. Regarding cases involving apparent intention consistently carried out by apparitions, she noted that, ". . . it is a rather remarkable fact that we have exceedingly little evidence in our collection clearly tending in this direction" (p. 99). This would seem to lessen the popular assumption that purposeful intention on the part of an apparition is a clear indication that the apparition represents the survival of a person's conscious awareness, an assumption upon which the authors' idea is apparently based (p. 155).

To account for RSPK, Jawer and Micozzi refer to the findings suggesting that RSPK has both a psychological aspect and an energetic aspect (Roll & Persinger, 1998). On this basis, they offer the suggestion that the RSPK agent ". . . becomes a generator of electromagnetic radiation, or at least a transducer for energy in the immediate environment" (p. 197). Moreover, they offer the radical suggestion that electromagnetism may be a medium of conveyance for repressed feeling and emotional information, as indicated by their conjecture that "Electromagnetic radiation . . . is *conducive to feeling*" (p. 197, their emphasis; see also pp. 405 and 441 for similar conjectures). However, aside from the RSPK case findings (which suggest that emotion and electromagnetism may be *correlated*, but not necessarily causally *connected* with each other) and neurological research that suggests that complex magnetic pulses applied to the medial temporal regions may artificially induce an emotion-related response (e.g., Richards et al., 1992) but not the other way around, there seems to be little (if any) clear evidence that emotion can be conveyed through electromagnetism.

On the other hand, there is an alternative suggestion that would require consideration of a third component to act as a sort of a mediator between the two, and that third component is PK. Here, as suggested by Roll and Persinger (1998, p. 193), PK would be utilized by the RSPK agent to modulate and focus magnetic energy in the environment rather than be a generator of it. In line with this view, a limited amount of PK research indicates deviations from nominal randomness in the output of random event generators (REGs) correlating with certain forms of emotional expression (Blasband, 2000, Lumsden-Cook, 2005a, 2005b). In addition, exploratory PK tests conducted by Puthoff and Targ (1975) with two psychics found evidence to suggest that the psychics were able to affect the wave patterns of magnetic fields, seemingly offering a weak hint that PK may affect external fields (although it must be recognized that it is not clear which target the psychics may have affected—the magnetic fields themselves, or the mechanical components of the magnetometers measuring the fields).

Elsewhere in the book, to further support their bodymind perspective, Jawer and Micozzi review the immunological and psychological evidence for apparent psychosomatic effects. In their attempt to relate these effects to psi, the authors cite organ transplant cases in which transplant recipients seem to spontaneously "recall" events in the lives of their donors (Dossey, 2008, Pearsall et al., 2002). The authors suggest that such cases may be accounted for through consideration of a type of "memory," which some have labeled "cellular memory" (op. cit.), that is somehow retained by the bodymind. However, they do not recognize the apparent similarity between these cases and those of psychometry, also known as token object reading (for a brief review of cases, see Roll, 2004). In psychometry, a psychic appears to receive psi-related impressions about a specific person or event through the handling of an inanimate physical object associated with that person or event. Similar to the organ transplant cases, psychometric impressions can be "memory"-like and correspond to verifiable events in the lives of living or deceased persons. This seems to suggest that the transplant cases may represent a form of psychometry. If that is the case, then it may be necessary to take into account psychometry cases involving *inanimate* (i.e. non-biological) objects as well as cases involving animate ones. It is unclear how Jawer and Micozzi's bodymind perspective would be able to do this in its current form.

In further attempts at extension to psi, Jawer and Micozzi cite the cases suggestive of reincarnation that have been extensively documented by the late Ian Stevenson and his associates at the University of Virginia, particularly those involving birthmarks that seem to correspond to injuries sustained by the previous personality (Stevenson, 1997; Tucker, 2005). To account for these cases, the authors suggest that repressed emotional energy resulting from traumatic experience is somehow transferred from one bodymind to another (p. 352). However, aside from anecdotal allusions to reports of "energy" produced by practitioners of mind–body disciplines (pp. 60–61), the authors provide little detail about emotional energy, how it may possibly be verified empirically, and the process by which it may be transferred in these cases.

In attempting to account for experiences of clairvoyance and precognition, Jawer and Micozzi suggest that these may be instances of the bodymind moving through space–time, offering the argument that: "When our emotions are being expressed (i.e. expanding into space), our experience of time should also be in some sense greater" (p. 374). The authors correctly point out that spontaneous ESP experiences tend to involve emotional events, although they do not seem to recognize that emotion may only be one part of ESP rather than the whole, as indicated by their assertion that one never hears of non-emotional ESP experiences (p. 374). While case studies do indicate that spontaneous ESP tends to involve (mostly negative) emotional events (e.g., Feather & Schmicker, 2005; Rhine, 1981; Stevenson, 1970), there are indeed also a small number of ESP experiences involving trivial (i.e. being of less acute emotional value) events. For instance, Louisa Rhine (1981) found that approximately 15% of her 2,878 ESP cases involved a trivial matter, and Ian Stevenson (1970) found that 9% of his 35 cases were trivial (see also his Table 4 for other case collections). One might also consider the early ESP tests by J. B. Rhine and colleagues using Zener cards, which are a relatively unemotional stimulus. Another example might be remote viewing of geographical locations (Targ & Puthoff, 1977), which can also be considered a relatively unemotional stimulus. While ESP experiences of trivial events do comprise only a small proportion of the overall database, their mere existence offers reason that there should be a way to account for them within Jawer and Micozzi's perspective. Again, it is unclear how this may be done in its current form.

In their attempt to account for out-of-body experiences (OBEs), Jawer and Micozzi seem to rely on nearly the same ideas they proposed for apparitions and RSPK, namely that emotional energy associated with the bodymind can be displaced into the environment through electromagnetism (p. 432). As noted previously, empirical support for these ideas appears to be severely limited at the present time. Largely on the basis of anecdotal accounts, the authors contend that reported OBEs may represent "an external aspect to the bodymind continuum" (p. 426), and that during such experiences, individuals may be outside of their bodyminds (p. 432), suggesting that they are in favor of an extrasomatic aspect to the OBE. While this may still be a possibility, they do not seem to fully recognize that OBE perceptions can appear to be indistinguishable either from vivid hallucinations or from ESP. For instance, the famed OBE experient Robert Monroe would often report having OBEs that were quite vivid and detailed, but not always accurate with regard to what he perceived while presumably out of body (e.g., see the tests by Tart, 1998, pp. 82–89). A similar observation was made by William Roll (1997, p. 55) regarding the last and clearest OBE he had personally experienced. During a study by Tart (1998, pp. 78–82), a female experient was able to accurately recall a random five-digit number following a brief OBE, which could possibly have been perceived through ESP. The authors do note (pp. 416–417) the studies conducted by the Psychical Research Foundation (Morris et al., 1978) in which a pet kitten belonging to Keith Harary seemed to calm at times when Harary attempted to visit it via OBE during randomly determined periods (which could possibly be accounted for by ESP). However, inconsistent or null results were observed in tests using other animal detectors, human detectors, and physical detectors, casting some doubt on the extrasomatic aspect. In the case of the latter, no definitive changes (including magnetic changes) were found whenever Harary attempted to visit a room where physical instrumentation was located. This would tend to refute Jawer and Micozzi's suggestion that electromagnetism may be involved in OBEs.

Despite the potential shortcomings with regard to their psi-related arguments, Jawer and Micozzi's book provides a well-written review of psychosomatic research that is accessible to the general reader. In addition, the authors present material of potential promise in Chapter 9. Guided by his ergonomic-related work in assessing the effect of air quality on the work performance of people inside office buildings, Jawer (2005) developed a questionnaire examining the degree to which people reporting apparitional experiences might be susceptible to certain forms of environmental illness (e.g., allergies, migraine headache, chemical sensitivity, depression, sensitivity to light or sound). To date, this questionnaire has been completed by 112 people, 62 of which have described themselves as being environmentally sensitive. Compared to a control group of 50 non-sensitives, the sensitive group reported a higher percentage of apparitional experiences (16% vs. 74%, respectively). Chapter 9 provides a detailed assessment of Jawer's survey results, which suggest that apparitional experients may be particularly sensitive to their surroundings. Currently the findings are mostly limited to those individuals who have had an apparitional experience, and the degree to which it can be further replicated and generalizable to individuals who have experienced other kinds of psi-related phenomena remains to be seen.

BRYAN J. WILLIAMS
University of New Mexico
Albuquerque, New Mexico
bwillliams74@hotmail.com

References

Blasband, R. A. (2000). The ordering of random events by emotional expression. *Journal of Scientific Exploration, 14,* 195–216.

Cannon, W. B. (1927). The James–Lange theory of emotions: A critical examination and an alternative. *American Journal of Physiology, 39,* 106–124.

Dossey, L. (2008). Transplants, cellular memory, and reincarnation. *Explore: The Journal of Science & Healing, 4,* 285–293.

Feather, S. R., & Schmicker, M. (2005). *The Gift: ESP, the Extraordinary Experiences of Ordinary People.* New York: St. Martin's Press.

James, W. (1884). What is emotion? *Mind, 9,* 188–205.

Jawer, M. (2005). Environmental sensitivity: A link with apparitional experience? *Proceedings of Presented Papers: The Parapsychological Association 48th Annual Convention* (pp. 71–85). Petaluma, CA: Parapsychological Association, Inc.

Lumsden-Cook, J. (2005a). Mind–matter and emotion. *Journal of the Society for Psychical Research, 69,* 1–17.

Lumsden-Cook, J. (2005b). Affect and random events: Examining the effects of induced emotion upon mind–matter interactions. *Journal of the Society for Psychical Research, 69,* 128–142.

Morris, R. L., Harary, S. B., Janis, J., Hartwell, J., & Roll, W. G. (1978). Studies of communication during out-of-body experiences. *Journal of the American Society for Psychical Research, 72,* 1–21.

Pearsall, P., Schwartz, G. E. R., & Russek, L. G. S. (2002). Changes in heart transplant recipients that parallel the personalities of their donors. *Journal of Near-Death Studies, 20,* 191–206.

Pert, C. B. (1997). *Molecules of Emotion: Why You Feel the Way You Feel.* New York: Scribner.

Puthoff, H., & Targ, R. (1975). Physics, entropy, and psychokinesis. In Oteri, L., (Ed.), *Proceedings of an International Conference: Quantum Physics and Parapsychology* (pp. 129–141). New York: Parapsychology Foundation, Inc.

Rhine, L. E. (1981). *The Invisible Picture: A Study of Psychic Experiences.* Metuchen, NJ: Scarecrow Press.

Richards, P. M., Koren, S. A., & Persinger, M. A. (1992). Experimental stimulation by burst-firing weak magnetic fields over the right temporal lobe may facilitate apprehension in women. *Perceptual and Motor Skills, 75,* 667–670.

Roll, W. G. (1982). The changing perspective on life after death. In: Krippner, S. (Ed.), *Advances in Parapsychological Research 3* (pp. 147–291). New York: Plenum Press.

Roll, W. G. (1997). My search for the soul. In: Tart, C. T., (Ed.), *Body Mind Spirit: Exploring the Parapsychology of Spirituality* (pp. 50–67). Charlottesville, VA: Hampton Roads Publishing.

Roll, W. G. (2004). Early studies on psychometry. *Journal of Scientific Exploration, 18,* 711–720.
Roll, W. G., & Persinger, M. A. (1998). Poltergeist and nonlocality: Energetic aspects of RSPK. *Proceedings of Presented Papers: The Parapsychological Association 41st Annual Convention* (pp. 184–198). Durham, NC: Parapsychological Association, Inc.
Sidgwick, E. M. [Mrs. H.] (1885). Notes on the evidence, collected by the Society, for phantasms of the dead. *Proceedings of the Society for Psychical Research, 3,* 69–150.
Stevenson, I. (1970). *Telepathic Impressions: A Review and Report of Thirty-Five New Cases.* Charlottesville, VA: University Press of Virginia.
Stevenson, I. (1997). *Where Reincarnation and Biology Intersect.* Westport, CT: Praeger.
Targ, R., & Puthoff, H. E. (1977). *Mind Reach: Scientists Look at Psychic Ability.* New York: Delacorte Press/Eleanor Friede.
Tart, C. T. (1998). Six studies of out-of-body experiences. *Journal of Near-Death Studies, 17,* 73–99.
Tucker, J. (2005). *Life before Life: A Scientific Investigation of Children's Memories of Previous Lives.* New York: St. Martin's Press.

La Connaissance Supranormale, Étude Expérimentale by Eugène Osty. Felix Alcan (Paris), 1922. 358 pp. Exergue, 2000. 358 pp. $47 (paperback). ISBN 9782911525377. [**Supernormal Faculties in Man: An Experimental Study** by Eugène Osty. Methuen & Co. (Paris), 1923. $250 (hardcover). ASIN B001OMVDEA.]

Eugène Osty was born in Paris on May 16th, 1874. Graduated as a Medical Doctor, he became interested in metapsychics after a dinner with a psychic who described his personality and those of several friends. From 1909 onward, he met with psychics in Paris to study their "lucidity". He published his first experimental results and analyses in 1913 in *Lucidité et Intuition* (Osty, 1913). Charles Richet, 1913 Nobel laureate for the Prize in Physiology or Medicine, invited Osty, in June 1914, to be part of the "dinner of the thirteen". Each 13th of the month for 20 years, Richet gathered French personalities interested in psychical research such as Henri Bergson and Camille Flammarion.

After the war, Osty published *Le Sens de la Vie Humaine* (Osty, 1919). He also joined the Institut Métapsychique International (IMI) and published in 1922 *La Connaissance Supranormale* (Osty, 1922/2000). He became director of IMI after Geley's accidental death in a plane crash and held this position until his death in 1938. While he had practiced as a doctor in Jouet-sur-l'Aubois since 1901, he gave up his activities as a physician from 1924 to 1931 to devote himself entirely to the study of psychic phenomena. As a metapsychist, he conducted many experiments with famous psychics such as Pascal Fortuny (Osty, 1926), but he also set up original and ingenious experiments, with his son, Marcel Osty, to test physical mediums such as Jean Guzik and Rudi Schneider. His last book about the topic was *Les Pouvoirs Inconnus de l'Esprit sur la Matière* (Osty & Osty, 1932). In addition to these books, he published more than 100 articles in psychical research journals.

In *La Connaissance Supranormale*, Osty described twelve years of research and analysis about "metagnomy", a term coined by Émile Boirac (Boirac, 1917) which means "beyond knowledge" and that can be seen as an equivalent term for psi. This book described different phenomenological aspects of metagnomy.

Osty started *La Connaissance Supranormale* with general remarks about mankind and science and discussed his ambition to elicit the interest of famous scientists with his book. He presented several examples, coming mainly from other researchers, about metagnomy in the medical domain. He reported in detail, for example, the case of an hysteric who described correctly to her doctor, under hypnosis, a part of a bone that she had in the intestine (p. 37). Osty concluded there was a transformation of kinesthetic perceptions into visual representations depending on the beliefs and naive representations of the internal body that the psychic had. He also described several of his own cases in which patients knew precisely the day, and even the hour, on which they would die even if they were not sick at the moment of the prediction (p. 44). Even more disturbing was the case of a young girl who said to her mother "you are putting death's socks on" a few hours before she was killed by the falling down of her house (p. 45).

Osty then proposed several striking cases of metagnomy about, for example, the death of Jean Jaurès (p. 58), and the way a knife used in a murder was found thanks to a psychic (p. 62). Using these different examples coming from literature, and his own observations, Osty argued that "the study of this faculty leads to the fact that subjects who possess it are so different in their own capacity, that there are not two of them who are similar" (p. 79). For this reason Osty insisted on the importance of taking into account the specificity of these abilities:

> Trying to control the reality of hyper-knowledge by an attempt to obtain one specific category of phenomena, without checking the type of subject that will be employed for it, is nonsense. Each subject who has the ability of surnormal knowledge is, for the researcher, as an instrument whose functioning and the conditions of utilization must be known before using it. (p. 82)

When the psychic was well-known, Osty considered that the "metagnome", as he called them, could be as useful for the doctor as a dog was for the hunter (p. 104). Osty also compared metagnomes to different microscope lenses, who had to be used with precision, because some of them were better able to describe general patterns while others were more prone to give accurate details (p. 142). He was also convinced that there is a link between psi and memory processes because metagnomes generally have a very good memory. Nevertheless, he also specified that "who can do more, cannot necessarily do less" (p. 79). For example, there is no sense asking a psychic to describe what is in an adjacent room under the pretext that he was able to precisely describe a personality.

Osty was convinced that psi subjects could be used in different settings if one considered precisely the ability of each metagnome and how to use it with efficiency. Besides medical diagnosis, in which, from his own experience, best results were obtained when doctors and metagnomes worked together, Osty thought they could be useful for the treatment of psychoneurosis (p. 105). He also proposed several successful examples of applications in other domains: a description of a child's personality in order to detect hidden abilities (p. 112), employment of a new domestic (p. 115), and even information to determine if a woman would be a good wife (p. 117)!

Osty was also convinced that metagnomy could be useful to know future events of his own life (he had seen many psychics for personal readings). He described in detail several cases of metagnomy and compared each sentence said by the psychic with what had been verified. One case was especially interesting: It began with general information, looking like simple cold reading processing (Hyman, 1989), and finished with accurate information such as the name and surname of several persons and precise personal events.

More generally, Osty wanted to see these abilities used in a more clever and efficient way:

> I dream of an epoch in which the period of ridiculous mysticism and skepticism will be closed, and in which metagnome subjects of good quality will be removed from their destiny of fortune-tellers, and will be selected, wisely educated, rationally prepared, and will become for men of science, finally skilled in this domain, psychic tools helping the experimental exploration of [the] latent transcend[ent] plan[s?] of [the] human being, and maybe of everything he lives. (p. 155)

But this dream was quite different from Osty's zeitgeist, and he criticized the attitude of several famous French psychologists concerning metapsychics, such as Binet and Vaschide (who, ironically, had his own year of death predicted rightly by a psychic working with Osty, Mme Fraya). Osty also criticized the supposed amalgam between neurosis and metagnomy, which he believed was the consequence of abusive generalization (p. 180).

From his detailed analysis of numerous sessions conducted with different psychics, Osty reached the conclusion that the "mind of the subject seemed able to communicate with all the individual elements of humankind if we give him sufficient support" (p. 195). This support was an "intermediate object" that enabled a link between the metagnome and the target, making the latter able to use a "creative power and knowledge beyond space and time" (p. 214). Nevertheless, this knowledge was characterized metaphorically most of the time. For example, some psychics gave true details about organs, but in a metaphoric manner and with a naive representation of the human body (p. 220). Osty thought that this phenomenon came from the fact that psi perceptions were a reconstructive hallucination produced by the mind of the metagnome.

Osty also discussed the great variability of results, as they can be "excellent to nothing" (p. 261), even with the same psychic. But the French researcher argued that good results did not depend only on the psychic: There could be distortions coming from the target, such as the person who was concerned by the reading (p. 262).

Nevertheless, the rate of errors was more frequent with nonselected subjects (p. 277), and several sessions with the same psychic may allow collection of more reliable information (p. 197). In Osty's view, this information can hardly concern global events, like, for example, the first world war from a general point of view, but Osty gave several examples in which the psychic had predicted the impact of the war when doing predictions for individuals. Consequently, these general events could be deduced from the comparison of individual readings.

Osty came to the conclusion that the study of errors and distortions of the psychic's descriptions was essential, and he examined them precisely in the last part of *La Connaissance Supranormale*. Indeed, if he had obtained "sessions ideally good", he never encountered "a perfect subject" (p. 319). He noted moreover that the metagnome not only caught the reality but also conveyed "fears, projects, desires and hopes" (p. 313).

Interestingly, even if Osty seemed quite critical about psychoanalysis, he had discovered the same processes of distortion as those described by Freud during the same period (Freud, 1932). Furthermore, other obstacles could lead to false conclusions concerning the descriptions of a psychic. Osty observed more particularly that "agreement between revelations of metagnome subjects can occur as much in error as in truth" (p. 328). He then presented a particularly long example in which up to ten different metagnomes gave a consistent description of a soldier who had disappeared. In the different descriptions, the soldier was supposed to be alive, and he was actually already dead. But the subjects gave the same details about what they thought had happened to the soldier in a long and coherent story as if all their descriptions were interconnected. Osty supposed that this effect came from a false interpretation of mental imagery by the psychics, and could also be the consequence of the desire of the person who asked for the reading (in this case, the father of the soldier, who of course hoped his son was still alive). More generally, in order to prevent errors and distortions, Osty advised experimenters to be careful with the type of questions they ask the metagnome even to the point of not letting him in on their "absolutely unknown" information (p. 341).

Osty concluded his book with the importance of proposing a hypothesis that could be experimentally tested. He showed great caution concerning the different interpretations of the observed phenomena and thought that future experiments would have to determine more generally the links between brain and mind. He finally apologized for giving only partial satisfaction to his scientific readers as he was only able to propose several paths that still had to be explored.

We can respect the modesty of a researcher who has proposed a very experimental and detailed approach using a qualitative methodology and which can be seen as a precursor of many later researches. First of all, Osty's observations about psi functioning influenced Warcollier's work about telepathy, which focused on images (Warcollier, 1921). This research was then a source of inspiration for the American government's remote viewing program known as "Stargate" (Hyman, 1996, Utts, 1996). From this point of view, Osty can be seen as one of the first to propose the way to carry out psi applications in an organized and detailed manner.

Thus, when we compare Osty's observations ninety years ago and what is currently proposed by remote viewers and researchers who have been working on psi applications (McMoneagle, 2000), we are surprised by the similarity of observations and recommendations. For example, Osty noted the importance of taking into account not only the mental state of the psychic but also the target, which can be associated with more recent findings concerning the importance of the nature of the target (May, Spottiswoode, & Faith, 2000). This similarity of descriptions and observations over time could be seen as a strong argument in favor of an uniform phenomenology of psi, especially when it is used for applications. Nevertheless, it is still necessary from our point of view to determine if this coherence is actually an astonishing cognitive illusion that has deluded several generations of scientists or if it is the consequence of a real form of perception. In any case, this kind of work has great interest from a phenomenological point of view.

More generally, we can also observe that only a few ideas about applications proposed by Osty have been exploited so far. For example, the use of metagnomes in psychotherapy and medical diagnosis has not been studied and evaluated seriously in large studies so far. This lack of psi applications is probably one of the consequences of the decrease and almost disappearance of an experimental and elitist approach such as the one that used to exist in the French metapsychic tradition. Current researchers should take more into account Osty's conceptions that it was absolutely necessary to select participants and to acknowledge their specificity in order to study and use their abilities. Interestingly, this question was a crucial point among scientists who have analyzed data from the Stargate program. This principle, absolutely essential in the elitist approach, could maybe explain current problems of reliability observed in universalist and quantitative research. Researchers should maybe use these criteria more, especially for applications, in order to produce original research.

Moreover, from a more classical psychological point of view, Osty has distinguished clearly pathological and nonpathological mental states associated with paranormal experiences as still described today by some researchers (Simmonds-Moore & Holt, 2007). On a more psychodynamic ground, Osty has also observed mechanisms of distortion in the unconscious as Freud (1932) did. Experimental work with psychics could still appear as an original way to understand unconscious processing, and we can regret that this path of research has not been explored so far.

Nevertheless, we can also deplore the lack of rigor, from our current scientific standards, of many experiments carried out by Osty. For example, in most of the readings, there are no double-blind conditions, and consequently it's difficult to determine in which way sensory cues could explain some of the results. Moreover, Osty gives us successful examples, but we also need to know what the rate of failure was.

Consequently, and as a whole, even if Osty's research can hardly be considered as proof of psi functioning, it represents a phenomenological approach that deserves careful and detailed attention. This kind of work is rare today, and we can still find inspiring remarks in this old book. For example, a detail mentioned by Osty has attracted my attention, being similar to my ideas and recent work with psychics (Rabeyron, 2008). Osty thus explained that:

> The increase in normal knowledge given to a metagnome subject increases in quantity and quality the supra-normal knowledge that the subject has access to. (p. 262)

This observation could have important consequences. We can see it more precisely in examples given by Osty. During several readings, the metagnome first gives very general information, that looks like, at first, a cold reading, but then gives accurate data such as a name and a description of personal events. Could we suppose that Osty obtained such surprising and precise results because the circumstances of the experiment were similar? And not because this setting would give access to normal clues, but because the access to classical information is a way to enhance the access to psi information? Roughly, it perhaps means there could be better psi interactions if there would be more classical interactions. This kind of analysis could also fit quite well with the

Model of Pragmatic Information (Lucadou, 1995) and Weak Quantum Theory (Lucadou, Römer, & Walach, 2007) and show how topical Osty's analyses still are.

THOMAS RABEYRON
Ph.D. Student, Clinical Psychology
Lyon II University and Edinburgh University
thomas.rabeyron@univ-lyon2.fr

References

Boirac, E. (1917). *L'avenir Des Sciences Psychiques*. Paris: Alcan.
Freud, S. (1932/1984). Rêve et occultisme. In *Nouvelles Conférences d'Introduction à la Psychanalyse*. Paris: Gallimard.
Hyman, R. (1989). *The Elusive Quarry: A Scientific Appraisal of Psychical Research*. New York: Prometheus Books.
Hyman, R. (1996). Evaluation of a program on anomalous mental phenomena. *Journal of Scientific Exploration, 10*(1), 31–58.
Lucadou, W. (1995). The model of pragmatic information (MPI). *European Journal of Parapsychology, 11,* 58–75.
Lucadou, W., Römer, H., & Walach, H. (2007). Synchronistic phenomena as entanglement correlations in generalized quantum theory. *Journal of Consciousness Studies, 14*(4), 50–74.
May, E., Spottiswoode, S., & Faith, L. (2000). The correlation of the gradient of Shannon entropy and anomalous cognition: Toward an AC sensory system. *Journal of Scientific Exploration, 14*(1), 53–72.
McMoneagle, J. (2000). *Remote Viewing Secrets: A Handbook*. Charlottesville: Hampton Roads Pub.
Osty, E. (1913). *Lucidité et Intuition: Étude Expérimentale*. Paris: Alcan.
Osty, E. (1919). *Le Sens de la Vie Humaine*. Paris: La Renaissance du Livre.
Osty, E. (1922/2000). *La Connaissance Supranormale: Étude Expérimentale*. Paris: Editions Exergue.
Osty, E. (1923). *Supernormal Faculties in Man: An Experimental Study*. London: Methuen & Co.
Osty, E. (1926). *Pascal Forthuny: Une Faculté de Connaissance Supra-Normale*. Paris: Alcan.
Osty, E., Osty, M. (1932). *Les Pouvoirs Inconnus de l'Esprit sur la Matière*. Paris: Alcan.
Rabeyron, T. (2008). *Analyse Cognitive des Expériences Exceptionnelles*. Ecole Normale Superieure de Cachan: Unpublished memory.
Simmonds-Moore, C., Holt, N. (2007). Trait, state and psi: A comparison of psi between clusters of scores on schizotypy in a ganzfeld and waking control condition. *Journal of the Society for Psychical Research, 71*(4), 197–215.
Utts, J. (1996). An assessment of the evidence for psychic functioning. *Journal of Scientific Exploration, 10*(1), 3–30.
Warcollier, R. (1921.) *La Télépathie—Recherches Expérimentales*. Paris: Alcan.

Phénomènes Psychiques au Moment de la Mort by Ernest Bozzano. Éditions de la B.P.S. (Paris), 1923. 261 pp. J. M. G. Editions (Agnières), 2001. 326 pp. €18.30. ISBN 2912507529.

Deathbed Visions: The Psychical Experiences of the Dying by Sir William F. Barrett. Methuen (London), 1926. 116 pp. Free at http://www.survivalafterdeath.org.uk/books/barrett/dbv/contents.htm. Aquarian Press (Wellingborough, Northamptonshire), 1986. 173 pp. £5.99 (paperback). ISBN 0850305209.

There is a long tradition of the association of "between death" and a variety of psychic phenomena, among them apparitions, physical phenomena, and what we refer to today as ESP. The writings of such authors as Gurney, Myers, and Podmore (1886), and Flammarion (1920–1922/1922–1923) are examples of this. The books reviewed here are important and influential representatives of this idea.

The first one was authored by Italian student of psychic phenomena Ernest Bozzano (1862–1943), who was well-known for his studies presenting numerous cases of psychic phenomena and for his strong defense of the idea of the survival of bodily death. In *Phénomènes Psychiques au Moment de la Mort,* Bozzano brought together three of his previously published studies about death-related phenomena, namely deathbed visions, music, and physical phenomena.

In the study of deathbed visions, Bozzano presented a classification consisting of visions of persons: (1) known to be dead and seen only by the dying individual; (2) not known to be dead and seen only by the dying individual; (3) seen both by the dying persons and by deathbed bystanders; (4) showing correspondences with information obtained through mediumistic communicators; (5) perceived only by relatives of the dying person located around or close to the deathbed; and (6) seen somewhat after death and in the same house where the dead body was located.

Bozzano presented examples and discussed those veridical visions in which the dying person perceived someone he or she did not know had died. He examined critically the idea that persons knowing about the death affected the dying individual via a subconscious telepathic message that produced a hallucination in the dying person. Bozzano objected to this explanation because he considered it unlikely that such communication would take place between individuals lacking affective rapport between them, a necessary precondition for telepathy, in his view. Furthermore, he believed telepathic transmission was unlikely because "in nearly all spontaneous telepathic phenomena the *agent* transmits to the *percipient* the hallucinatory vision of their own person, and not that of another person . . . ." (p. 51) (this, and other translations, are mine).

Referring to subjective hallucinations, Bozzano wrote that "if the phenomena in question have as a cause the thoughts of the moribund . . . the dying person . . . should perceive more frequently hallucinatory forms representing living persons" (p. 109), something he said did not take place. However, Bozzano could have been dealing with a biased sample of cases. We must remember that many of his sources were spiritualistic and psychical research books and journals. It is unlikely that the authors and editors of such publications would have been interested in accounts of visions of the living, unless they were veridical visions.

Bozzano also mentioned a case in which a man saw apparitions at his wife's deathbed around her dying body and her "astral body" floating above her physical body. He considered the latter an objective "fluidic doubling." Interestingly, and because one of the apparitions seen was of a woman in a Greek costume and with a crown on her head, Bozzano speculated on the possibility of a "telepathic–symbolic projection" (p. 103) from a spiritual entity.

The section about physical phenomena involved various events corresponding to deaths. As I have mentioned elsewhere (Alvarado, 2006:135), out of 13 accounts presented by Bozzano, the effects consisted of: falling objects (54%), clocks stopping or starting (23%), objects rocking or shaking (8%), objects breaking or exploding (8%), and lights turning on or off (8%). Bozzano argued that cases in which the dying person and the physical event were distant from each other showed that the effect was not physical, but had to be psychical. This suggested to him the presence of the spirit of a dead person at the location in which the event took place. Furthermore, he noticed that some cases involved intention.

In the third part of the book, Bozzano discussed what he called "transcendental music." He presented cases that took place at deathbeds and after deaths. But Bozzano also discussed mediums who performed with musical instruments, telepathically perceived music, and music heard in hauntings.

The following is an example of a case of music cited by Bozzano (pp. 230–231) related to an apparition, which I take from the original:

> In October, 1879, I was staying at Bishopthorpe, near York, with the Archbishop of York. I was sleeping with Miss Z. T., when I suddenly saw a white figure fly through the room from the door to the window. It was only a shadowy form and passed in a moment. I felt utterly terrified, and called out at once, "Did you see that ?" and at the same time Miss Z. T. exclaimed, "Did you *hear* that?" Then, I said, instantly, "I saw an angel fly through the room," and she said, "I heard an angel singing" (Sidgwick et al., 1894:317–318).

Bozzano argued that the case represented "two simultaneous supernormal manifestations that, due to the special idiosyncracies of the percipients, were perceived separately" (p. 231).

Collective percipience of music, Bozzano argued, eliminated suggestion and hallucination as an explanation. In many of the cases the dying person "did not participate in the collective hearing of transcendental music, which excludes all possibility of explaining the facts by assuming a hallucination having its origin in the mentality of the dying person" (pp. 258–259). This referred to the idea that the dying person affected bystanders via a process of telepathic transmission.

The analyses of these cases, and of other psychic phenomena, led Bozzano to argue that he had found proof for survival of death. This proof, he argued, came from different lines of evidence and types of cases that, when considered together, supported each other.

Some years after Bozzano's book appeared, William Fletcher Barrett's (1844–1925) *Deathbed Visions* was published, a book that has long been recognized as a classic on the subject. Unfortunately, this is an incomplete study because its author died before he could finish it. The chapters presented here were put together by the author's wife, physician Florence E. Barrett, who decided not to add anything to the book so as to keep the author's thought intact.

Barrett was a physicist with a lifelong interest in psychic phenomena. He was a founding member and one of the first vice-presidents of the Society for Psychical Research in 1882, and served in later years as a council member and as President of the Society. Barrett published on such varied topics as telepathy, mediumship, mesmerism, and dowsing. A believer in the nonphysical nature of the mind, Barrett wrote in an autobiographical essay, "psychical research will demonstrate to the educated world, not only the existence of a *soul in man, but also the existence of a soul in Nature* . . . ." (Barrett, 1924:296).

Barrett summarized his outlook in the first chapter as follows:

> It is well known that there are many remarkable instances where a dying person, shortly before his or her transition from the earth, appears to see and recognize some deceased relatives or friends. We must, however, remember the fact that hallucinations of the dying are not very infrequent. Nevertheless, there are instances where the dying person was *unaware* of the previous death of the spirit form he sees, and is therefore astonished to find in the vision of his or her deceased relative one whom the percipient believes to be still on earth. These cases form, perhaps, one of the most cogent arguments for survival after death, as the evidential value and veridical (truth telling) character of these Visions of the Dying is greatly enhanced when the fact is undeniably established that the dying person was wholly ignorant of the decease of the person he or she so vividly sees. (p. 1)

Barrett included in the second chapter several cases about visions corresponding to people not known to be dead at the time of the vision. He took the following case from James H. Hyslop, who in turn took the account from Minot Savage. The case reads as follows:

> In a neighbouring city were two little girls, Jennie and Edith, one about eight years of age and the other but a little older. They were schoolmates and intimate friends. In June, 1889, both were taken ill of diphtheria. At noon on Wednesday Jennie died. Then the parents of Edith, and her physician as well, took particular pains to keep from her the fact that her little playmate was gone. They feared the effect of the knowledge on her own condition. To prove that they succeeded and that she did not know, it may be mentioned that on Saturday, June 8th, at noon, just before she became unconscious of all that was passing about her, she selected two of her photographs to be sent to Jennie, and also told her attendants to bid her good-bye.
>
> She died at half-past six o'clock on the evening of Saturday, June 8th. She had roused and bidden her friends good-bye, and was talking of dying, and seemed to have no fear. She appeared to see one and another of the friends she knew were dead. So far it was like other similar cases. But now suddenly, and with every appearance of surprise, she turned to her father and exclaimed, "Why, papa, I am going to take Jennie with me!" Then she added, "Why, papa! you did not tell me that Jennie was here!" And immediately she reached out her arms as if in welcome, and said, "Oh, Jennie, I'm so glad you are here!" (Barrett, pp. 18–19)

Later chapters included such fascinating phenomena as apparitions seen by persons around the deathbed, visions of distant events, music heard by the dying person or by bystanders, and visions of what some described as the separation of the spirit from the physical body at death. Clearly the content of the book was not limited to cases of visions of the dying.

While the book consists mainly of case reports, on occasion Barrett discussed explanations for them. For example, in a case in which two sisters saw the faces of their two dead brothers looking at their dying sister, he mentioned Frank Podmore's speculation that the image was created by telepathy from the dying sister. Barrett wrote that "this explanation is less tenable and quite as unlikely as is the percipience of spirit forms by the dying person and sometimes by those present" (p. 75).

It is obvious that both Bozzano and Barrett were influenced by the work of previous persons, as seen in their citations of works, many of which come from the spiritualistic and psychical research literatures. Regarding deathbed visions, some of their predecessors were Frances Power Cobbe (1877) and James H. Hyslop (1907). Barrett's death prevented him from using in more detail the work of Bozzano. According to Barrett's wife, her husband had marked parts of Bozzano's book reviewed here. She wrote:

> He was specially interested in Bozzano's observation that if the phenomena were caused by the thoughts of the dying person being directed to those he loved, the appearances might be expected to represent living persons at least as frequently as deceased persons who had long passed from this world, whereas no records had come to hand of dying persons seeing at their bedside visions of friends still living. (Barrett, pp. vii–viii)

While Barrett understandably did not provide much analysis, Bozzano did. However, his conclusions sometimes were too definitive. Certainly they depended on theoretical assumptions that could be questioned, such as the way telepathy manifests. In later years Bozzano continued to make similar arguments in favor of survivalistic interpretations. His last statements were made posthumously in his books *Musica Transcendentale* (1943/1982), *Le Visioni dei Morenti* (1947), and *La Psiche Domina la Materia* (1948), which included new cases.

There is no question that research on these topics has been neglected (Alvarado, 2006; for an exception see Brayne, Lovelace, & Fenwick, 2008). Leaving aside the general topic of apparitions of the dead, we should mention the deathbed visions work of Karlis Osis (1961) and of Osis and Haraldsson (1997), the most sophisticated work on the subject conducted to date. Some work has been conducted with death-related physical phenomena (Piccinini & Rinaldi, 1990, Rhine, 1963) and music (Rogo, 1970, 1972). Other topics, such as collectively perceived deathbed cases and the cases of emanations from the dying body, have received much less attention (e.g., Crookall, 1967). Unfortunately, and with the exception of the above-mentioned research, the study of the phenomena outlined by Bozzano and Barrett has not received systematic attention.

Both Bozzano and Barrett performed a service for later researchers by presenting an organized catalog of observations. To this day individuals interested in the phenomena they discuss find useful illustrative cases in their books. But their contribution was not limited to this. They also documented the variety of death-related phenomena, something that has also been done by other authors, such as Flammarion (1920–1922/1922–1923). Furthermore, reading through the books reviewed here modern readers can get a good idea of the features of these experiences. Another contribution is that these studies also remind us of the important conceptual issues underlying these phenomena, particularly the issue of survival of bodily death.

It is to be hoped that new interest in these phenomena goes beyond popular discussions (e.g., Wills–Brandon, 2000), and beyond purely descriptive studies that are limited to case presentations, as seen in some of the literature available today about "after-death" manifestations (e.g., Guggenheim & Guggenhein, 1995/1997). As I have argued elsewhere in terms of selected near-death phenomena, much remains to be done in this area, considering such aspects as prevalence, the features of the experiences, the characteristics of the experiencers, the relationship of the phenomena to other variables, and hypothesis testing (Alvarado, 2006). But future attempts to develop new research in this area will benefit from attention to Bozzano, Barrett, and other pioneers.

CARLOS S. ALVARADO
Atlantic University
215 67th Street, Virginia Beach, VA, 23451
carlos.alvarado@atlanticuniv.edu

References

Alvarado, C. S. (2006). Neglected near-death phenomena. *Journal of Near-Death Studies, 24,* 131–151.
Barrett, W. F. (1924). Some reminiscences of fifty years' psychical research. *Proceedings of the Society for Psychical Research, 34,* 275–297.
Bozzano, E. (1947). *Le Visioni dei Morenti.* Verona: Europa.
Bozzano, E. (1948). *La Psiche Domina la Materia: Dei Fenomeni di Telekinesia in Rapporto con Eventi di Morte.* Verona: Europa.
Bozzano, E. (1982). *Musica Transcendentale.* Rome: Mediterrance. (Original work published 1943)
Brayne, S., Lovelace, H., Fenwick, P. (2008). End-of-life experiences and the dying process in a Gloucestershire nursing home as reported by nurses and care assistants. *American Journal of Hospice and Palliative Medicine, 25,* 195–206.
Cobbe, F. P. (1877). The peak in Darien: The riddle of death. *Littell's Living Age, 19*(s.5), 374–379.
Crookall, R. (1967). *Events on the Threshold of the After Life.* Moradabad, India: Darshana International.
Flammarion, C. (1922–1923). *Death and Its Mystery* (3 Vols.). New York: Century. (First published in French, 1920–1922)
Guggenheim, B., & Guggenheim, J. (1997). *Hello from Heaven.* New York: Bantam Books. (Original work published 1995)
Gurney, E., Myers, F. W. H., & Podmore, F. (1886). *Phantasms of the Living* (2 vols.). London: Trübner.
Hyslop, J. H. (1907). Visions of the dying. *Journal of the American Society for Psychical Research, 1,* 45–55.
Osis, K. (1961). *Deathbed Observations by Physicians and Nurses* (Parapsychological Monographs No. 3). New York: Parapsychology Foundation.
Osis, K., Haraldsson, E. (1997). *What They Saw . . . at the Hour of Death* (3rd ed.). Norwalk, CT: Hastings House.
Piccinini, G., Rinaldi, G. M. (1990). *I Fantasmi dei Morenti: Inchiesta su una Credenza.* Viareggio, Italy: Il Cardo.
Rhine, L. E. (1963). Spontaneous physical effects and the psi process. *Journal of Parapsychology, 27,* 84–122.
Rogo, D. S. (1970). *NAD: A Study of Some Unusual "Other World" Experiences.* New York: University Books.

Rogo, D. S. (1972). *A Psychic Study of the "Music of the Spheres" (NAD Volume II)*. Secaucus, NJ: University Books.
Sidgwick, H., Johnson, A., Myers, F. W. H., Podmore, F., & Sidgwick, E. M. (1894). Report on the Census of Hallucinations. *Proceedings of the Society for Psychical Research, 10*, 25–422.
Wills-Brandon, C. (2000). *One Last Hug before I Go: The Mystery and Meaning of Deathbed Visions*. Deerfield Beach, FL: Health Communications.

Allan Kardec und der Spiritismus in Lyon um 1900. Geisterkommunikation als Soziales Phänomen by Katrin Heuser. VDM Verlag Dr. Müller, 2008. 120 pp. €59 (paperback). ISBN 9783639072587.

Judging from the content, structure, and layout of Heuser's micro-study of French spiritism in Lyon c. 1900, which is distributed by VDM (a German publisher specialising in academic theses), the book appears to be the published but self-edited version of the author's *Magisterarbeit,* or M.A. thesis, in cultural studies, though background information regarding the genesis of the book is entirely lacking. While the back cover blurb announces that the book is intended for readers interested in the science, sociology, and historical roots of spiritism, it is only the second (though dependent on the third) aspect that Heuser's study addresses adequately.

The study is based on the activities of the spiritist societies *Les Indépendants Lyonnais* (founded in 1890) and the *Société spirite pour l'Oeuvre de la Crèche* (founded in 1904) in Lyon, the historical capital of French spiritism or Kardecism. Using primary sources such as membership lists, police records, and the groups' periodicals and pamphlets, the author investigates the personal backgrounds of the founders, propagandists, and general members of the two groups, their social structures and aims, and the strictness of adherence to Kardec's original doctrines in relation to the groups' specific social interests. In a brief excursion, Heuser compares the spiritist scene of Lyon to that of the German capital of spiritism (or spiritualism), *fin-de-siècle* Leipzig. The theoretical framework for Heuser's historical study is Berger and Luckmann's social constructivist model of knowledge and reality. Contrary to previous authors' writing on the social and cultural history of French spiritism, such as Laplantine and Aubrée, Bergé, and Sharp, Heuser finds that social class did not determine involvement in spiritist societies, and that gender and biographical factors were more reliable determinants—at least for her small Lyon sample.

ALLAN KARDEC

Owing to the nature of the study as a work in cultural or social history rather than as history of science—but contradicting the misleading announcement in the book blurb—the science of Lyonnaise spiritism is not discussed at all. Although the author finds that the groups she investigated greatly differed in terms of scientific approaches to spiritism (i.e. *Les Indépendants* was supposed to have a strong emphasis on experimental seances and purported having scientific evidence for postmortem survival in general, while the *Société spirite* was almost entirely concerned with social welfare), we learn nothing about the methodology and scientific rigor (or possible lack thereof) employed in the experimental seances alluded to.

Also regrettably, the author's expertise in the general history of spiritualism seems somewhat wanting. For example, while Heuser rightly claims that the Hydesville incidents around the Fox sisters in 1848 gave birth to modern spiritualism as a large-scale movement, she makes the false conjecture that in Hydesville "for the first time, raps were interpreted as the manifestation of spirits" (p. 3, my translation). After all, the very coinage and use of the German word *poltergeist* ("rumbling spirit") precedes modern spiritualism by at least three centuries (see also, e.g., Kiesewetter, 1886, Gauld & Cornell, 1979). While the author draws upon works of important French spiritists such as Gabriel Delanne, Léon Denis, and Kardec himself, the only non-French "insider" history of spiritualism referred to (through its French edition) is Arthur Conan Doyle's (Conan misspelled as *Canon* throughout) notoriously unreliable *History of Spiritualism*.

Considering the book's unprofessional layout, stylistic flaws, and lack of index, VDM appears to be a provider of "quick and dirty" publishing-on-demand rather than a traditional academic publisher. To ensure additional academic quality control as well as a broader reception of her results, rather than hiding them away in this apparently self-edited and rather overpriced booklet, the author would do well to attempt boiling down her thesis and then submitting its main findings to a professional cultural studies journal.

ANDREAS SOMMER
Department of Science and Technology Studies
University College London
London, United Kingdom
a.sommer@ucl.ac.uk

References

Gauld, A., & Cornell, A. D. (1979). *Poltergeists*. London: Routledge & Kegan Paul.
Kiesewetter, C. (1886). Zur Vorgeschichte des modernen "Geisterklopfens". *Sphinx, 1*, 213–215.

Further Book of Note

El Mundo Oculto de Los Sueños. Metáfora y Significado para Comprender Toda Su Riqueza [The Hidden World of Dreams: Metaphor and Meaning to Understand All Your Wealth] by Alejandro Parra. Libros Aula Magna, 2009. 320 pp. $25 (paperback). ISBN 9789501742510.

Alejandro Parra begins with an overview of theories that have guided many modern researchers, and many not-so-modern, for he provides historical accounts of their most hard-won and valuable accomplishments. Insightful historical references make Parra's exposition informative for multiple topics, giving the reader the experience of journeying from their interpersonal experiences, memories, and even fantasies, to scientific achievements, ancient cultural traditions, and an unsuspected social awareness.

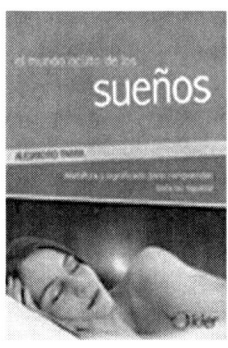

Parra respects past traditions that held magic and mysticism in strong regard, for throughout the book a clear sense of both awe and informed instruction is clearly felt by the selection not only of interpretative options and dreamwork based on different therapeutic approaches, but also their creative and personal potential, including the more fantastic interactions and qualities some of us may share. But, just as in life where it cannot be all peaches and cream, nightmares and traumatic and unpleasant dreams also are discussed.

The author's work here shows his sense of identification with the subject matter and his many years of work in the field, and expands on his previous book *Sueños: Cómo Interpretar Sus Mensajes* published by Kier Editorial. Just as it takes the reader on a self-journey, the book also plays ambassador to Argentina's exotic night life, giving detailed accounts of statistics taken from an online survey of more than 2,600 dreamers. And the book also makes it possible for readers to share their experiences in an ever-growing body of work, through a dream imagery questionnaire. This work is highly pragmatic for both layperson and professional.

cr. Carlos Adrián Hernández Tavares

Stanley Krippner
*Alan Watts Professor of Psychology
Saybrook University, San Francisco, CA*

SSE NEWS

Announcement for 2010 Grant

The Helene Reeder Memorial Fund for Research into Life after Death (HRF)

The Helene Reeder Fund is pleased to announce the availability of grants for small and medium-sized scientific research projects concerning the issue of Life after Death. Grants will be awarded in the range of €500–5,000 maximum. The topic Research into Life after Death should constitute the main objective of the project. Applications in English are to be submitted by email to the HRF to edgar.muller@comhem.se and should include:

- detailed description of the project, including the objectives of the project
- methodology
- cost budget
- timetable
- plans to publish the results in scientific journals
- CV of the applicant
- how the applicant plans to report back to the HRF about progress and results
- any other financing than from HRF

Applications should be received not later than October 30, 2010. It is the intention of the HRF to evaluate the applications and to make decisions regarding the grants before the end of December. Applicants will be notified by email after the decisions are made and the grants will be payable during December. For further information, please apply to edgar.muller@comhem.se

— Stockholm, March 2010

30th Annual SSE Meeting

Boulder, Colorado, June 21–23, 2011
Topics so far: Sociology of Science; Archaeological Mysteries

www.ScientificExploration.org

Society for Scientific Exploration

Executive Committee

Dr. Bill Bengston
SSE President
St. Joseph's College
Patchogue, New York

Professor Robert G. Jahn
SSE Vice-President
School of Engineering and Applied Science
Princeton University
Princeton, New Jersey

Dr. Mark Urban-Lurain
SSE Secretary
College of Natural Science
Michigan State University
111 N. Kedzie Lab
East Lansing, Michigan 48824

Dr. John Reed
SSE Treasurer
John Hopkins University
P. O. Box 16347
Baltimore, Maryland 21210

Ms. Brenda Dunne
SSE Education Officer
International Consciousness Research
    Laboratories
Princeton, New Jersey

Professor Garret Moddel
SSE Past President
Department of Electrical and Computer
    Engineering
University of Colorado at Boulder
Boulder, Colorado

Council

Dr. John B. Alexander
United States Army (Retired)
The Apollinaire Group
Las Vegas, Nevada

Dr. Richard Blasband
Center for Functional Research
Sausalito, California

Dr. Courtney Brown
Emory University
Atlanta, Georgia

Dr. Bernard Haisch
California Institute for Physics & Astrophysics
Redwood Shores, California

Dr. Roger Nelson
Global Consciousness Project
Princeton University
Princeton, New Jersey

Dr. Glen Rein
Ridgway, Colorado

Dr. Dominique Surel
Evergreen, Colorado

Dr. Chantal Toporow
Northrup Grumman
Redondo Beach, California

Appointed Officers

Professor Peter A. Sturrock
SSE President Emeritus and Founder
Stanford University
Stanford, California

Professor Charles Tolbert
SSE President Emeritus
Department of Astronomy
University of Virginia
Charlottesville, Virginia

Professor Lawrence Fredrick
SSE Secretary Emeritus
University of Virginia
Charlottesville, Virginia

P. David Moncrief
JSE Book Review Editor
Memphis, Tennessee

L. David Leiter
Associate Members' Representative
Willow Grove, Pennsylvania

Erling P. Strand
European Members' Representative
Østfold College
Halden, Norway

Professor Henry Bauer
JSE Editor Emeritus
Dean Emeritus, Virginia Tech
Blacksburg, Virginia

Index of Previous Articles in the *Journal of Scientific Exploration*

Vol: No	Article	Author(s)
1:1	A Brief History of the Society for Scientific Exploration	P. Sturrock
	Aterations in Recollection of Unusual and Unexpected Events	D. Hall et al.
	Toward a Quantitative Theory of Intellectual Discovery (Esp. in Phys.)	R. Fowler
	Engineering Anomalies Research	R. Jahn et al.
	Common Knowledge about the Loch Ness Monster	H. Bauer
	An Analysis of the Condon Report on the Colorado UFO Project	P. Sturrock
1:2	The Strange Properties of Psychokinesis	H. Schmidt
	What Do We Mean by "Scientific?"	H. Bauer
	Analysis of a UFO Photograph	R. Haines
	Periodically Flashing Lights Filmed off the Coast of New Zealand	B. Maccabee
2:1	Commonalities in Arguments over Anomalies	H. Bauer
	Remote Viewing and Computer Communications—An Experiment	J. Vallee
	Is There a Mars Effect?	M. Gauquelin
	Raising the Hurdle for the Athletes' Mars Effect	S. Ertel
2:2	UFOs and NASA	R. Henry
	The Nature of Time	Y. Terzian
	Operator-Related Anomalies in a Random Mechanical Cascade	B. Dunne et al.
	Evidence for a Short-Period Internal Clock in Humans	T. Slanger
	Three New Cases of Reincarnation Types in Sri Lanka with Written Records	I. Stevenson et al.
3:1	Arguments Over Anomalies: H. \?Polemics	H. Bauer
	Anomalies: Analysis and Aesthetics	R. Jahn
	Trends in the Study of Out-of-Body Experiences	C. Alvarado
	A Methodology for the Objective Study of Transpersonal Imagery	W. Braud/ M. Schlitz
	The Influence of Intention on Random and Pseudorandom Events	D. Radin/J. Utts
	Case of Possession Type in India with Evidence of Paranormal Knowledge	I. Stevenson et al.
3:2	New Ideas in Science	T. Gold
	Photo Analysis of an Aerial Disc Over Costa Rica	R. Haines/J. Vallee
	Three Cases of Children in Northern India Who Remember a Previous Life	A. Mills
	"Signatures" in Anomalous Human–Machine Interaction Data	D. Radin
	A Case of Severe Birth Defects Possibly Due to Cursing	I. Stevenson
4:1	Biochemical Traumatology/Plant Metabolic Disorders in a UFO Landing	M. Bounias
	Return to Trans-en-Provence	J. Vallee
	Analysis of Anomalous Physical Traces: 1981 Trans-en-Provence UFO Case	J. Velasco
	Physical Interpretation of Very Small Concentrations	H. Bauer
	Luminous Phenomena and Seismic Energy in the Central United States	J. Derr/ M. Persinger
	Photo Analysis of an Aerial Disc Over Costa Rica: New Evidence	R. Haines/J. Vallee
	A Scientific Inquiry into the Validity of Astrology	J. McGrew/ R. McFall
	Planetary Influences on Human Behavior: Absurd for a Scientific Explanation?	A. Müller
	Five Arguments against Extraterrestrial Origin of Unidentified Flying Objects	J. Vallee
4:2	Using the Study of Anomalies To Enhance Critical Thinking in the Classroom	M. Swords
	Observations of Electromagnetic Signals Prior to California Earthquakes	M. Adams
	Bayesian Analysis of Random Event Generator Data	W. Jefferys
	Moslem Case of Reincarnation Type in Northern India: Analysis of 26 Cases	A. Mills
	Electromagnetic Disturbances Associated with Earthquakes	M. Parrot
	Extrasensory Interactions between Homo Sapiens and Microbes	C. Pleass/N. Dey
	Correlation between Mental Processes and External Random Events	H. Schmidt
	Phobias in Children Who Claim To Remember Previous Lives	I. Stevenson
	A Gas Discharge Device for Investigating Focused Human Attention	W. Tiller

	Radio Emissions from an Earthquake	J. Warwick
5:1	The Cydonian Hypothesis	J. Brandenburg et al.
	Cases in Burma, Thailand, and Turkey: Aspects of I. Stevenson's Research	J. Keil
	Effects of Consciousness on the Fall of Dice: A Meta-Analysis	D. Radin/D. Ferrari
	The Wasgo or Sisiutl: A Cryptozoological Sea-Animal	M. Swords
	The Extraterrestrial Hypothesis Is Not That Bad	R. Wood
	Toward a Second-Degree Extraterrestrial Theory of UFOs	J. Vallee
	Low-Frequency Emissions: Earthquakes and Volcanic Eruptions in Japan	T. Yoshino
5:2	Eccles's Model of Mind–Brain Interaction and Psychokinesis	W. Giroldini
	Ball Lightning and St. Elmo's Fire as Forms of Thunderstorm Activity	A. Grigor'ev et al.
	Social Scientific Paradigms for Investigating Anomalous Experience	J. McClenon
	Count Population Profiles in Engineering Anomalies Experiments	R. Jahn et al.
	Children Claiming Past-Life Memories: Four Cases in Sri Lanka	E. Haraldsson
6:1	Can the UFO Extraterrestrial Hypothesis and Vallee Hypotheses Be Reconciled?	W. Bramley
	Learning for Discovery: Establishing the Foundations	R. Domaingue
	On the Bayesian Analysis of REG Data (Response from W. Jefferys)	Y. Dobyns
	Electrodynamic Activities and Their Role in the Organization of Body Pattern	M. W. Ho et al.
6:2	Review of Approaches to the Study of Spontaneous Psi Experiences	R. White
	Survival or Super-Psi?: Interchange Responses	I. Stevenson/S. Braude
	The Psychokinesis Effect: Geomagnetic Influence, Age and Sex Differences	L. Gissurarson
	Are Reincarnation Type Cases Shaped by Parental Guidance?	S. Pasricha
6:3	Heim's Theory of Elementary Particle Structures	T. Auerbach
	Better Blood through Chemistry: A Laboratory Replication of a Miracle	M. Epstein
	The Gauquelin Effect Explained? Comments on Müller's Planetary Correlations	S. Ertel
	The Gauquelin Effect Explained? A Rejoinder to Ertel's Critique	A. Müller
	Ball Lightning Penetration into Closed Rooms: 43 Eyewitness Accounts	A. Grivor'ev et al.
	A Series of Possibly Paranormal Recurrent Dreams	I. Stevenson
6:4	Experiments in Remote Human/Machine Interaction	B. Dunne et al.
	A Low Light Level Diffraction Experiment for Anomalies Research	S. Jeffers et al.
	A New Look at Maternal Impressions: An Analysis of 50 Published Cases	I. Stevenson
	Alternative Healing Therapy on Regeneration Rate of Salamander Forelimbs	D. Wirth et al.
7:1	Accultured Topographical Effects of Shamanic Trance Consciousness	P. Devereux
	Mainstream Sciences vs. Parasciences: Toward an Old Dualism?	G. L. Eberlein
	Existence of Life and Homeostasis in an Atmospheric Environment	S. Moriyama
	A Guide to UFO Research	M. D. Swords
7:2	Non-Causality as the Earmark of Psi	H. Schmidt
	Adequate Epistemology for Scientific Exploration of Consciousness	W. W. Harman
	Puzzling Eminence Effects Might Make Good Sense	S. Ertel
	Comments on Puzzling Eminence Effects	J. W. Nienhuys
	A Systematic Survey of Near-Death Experiences in South India	S. Pasricha
	The Willamette Pass Oregon UFO Photo Revisited: An Explanation	I. Wieder
7:3	Near Death Experiences: Evidence for Life After Death?	M. Schröter-Kunhardt
	Analysis of the May 18, 1992, UFO Sighting in Gulf Breeze, Florida	B. Maccabee
	Selection Versus Influence in Remote REG Anomalies	Y. Dobyns
	Dutch Investigation of the Gauquelin Mars Effect	J. Nienhuys
	Comments on Dutch Investigations of the Gauquelin Mars Effect	S. Ertel
	What Are Subtle Energies?	W. Tiller
7:4	Explaining the Mysterious Sounds Produced by Very Large Meteor Fireballs	C. S. L. Keay
	Neural Network Analyses of Consciousness-Related Patterns	D. I. Radin
	Applied Parapsychology: Studies of Psychics and Healers	S. A. Schouten
	Birthmarks and Birth Defects Corresponding to Wounds on Deceased Persons	I. Stevenson

	The "Enemies" of Parapsychology	R. McConnell
8:1	Survey of the American Astronomical Society Concerning UFOs: Part 1	P. Sturrock
	Anatomy of a Hoax: The Philadelphia Experiment Fifty Years Later	J. Vallee
	Healing and the Mind: Is There a Dark Side?	L. Dossey
	Alleged Experiences Inside UFOs: An Analysis of Abduction Reports	V. Ballester Olmos
	What I See When I Close My Eyes	R. Targ
8:2	Survey of the American Astronomical Society Concerning UFOs: Part 2	P. Sturrock
	Series Position Effects in Random Event Generator Experiments	B. Dunne et al.
	Re-Examination of the Law of Conservation of Mass in Chemical Reactions	K. Volkamer et al.
	The 'Genius Hypothesis': Exploratory Concepts for Creativity	E. Laszlo
8:3	Survey of the American Astronomical Society Concerning UFOs: Part 3	P. Sturrock
	Strong Magnetic Field Detected Following a Sighting of an UFO	B. Maccabee
	Complementary Healing Therapy for Patients with Type I Diabetes Mellitus	D. P. Wirth
	Report of an Indian Swami Claiming to Materialize Objects	E. Haraldsson
8:4	Scientific Analysis of Four Photos of a Flying Disk Near Lac Chauvet, France	Pierre Guérin
	A Linear Pendulum Experiment: Operator Intention on Damping Rate	R. D. Nelson
	Applied Scientific Inference	P. A. Sturrock
	The Mind-Brain Problem	J. Beloff
9:1	Unconventional Water Detection: Field Test of Dowsing in Dry Zones: Part 1	H. Betz
	Digital Video Analysis of Anomalous Space Objects	M. Carlotto
	The Critical Role of Analytical Science in the Study of Anomalies	M. Epstein
	Near-Death Experiences in South India: A Systematic Survey	S. Pasricha
	Human Consciousness Influence on Water Structure	L. Pyatnitsky/ V. Fonkin
9:2	Unconventional Water Detection: Field Test of Dowsing in Dry Zones: Part 2	H. Betz
	Semi-molten Meteoric Iron Associated with a Crop Formation	W. Levengood/MJ. Burke
	Experiments on a Possible g-Ray Emission Caused by a Chemical Process	V. Noninski et al.
	The Effect of Paranormal Healing on Tumor Growth	F. Snel/ P. van der Sijde
	Psychokinetic Action of Young Chicks on the Path of an Illuminated Source	R. Peoc'h
	Eddington's Thinking on the Relation between Science and Religion	A. Batten
	Two Kinds of Knowledge: Maps and Stories	H. Bauer
9:3	Experiments on Claimed Beta Particle Emission Decay	V. Noninski et al.
	Assessing Commonalities in Randomly Paired Individuals	T. Rowe et al.
	Anomalously Large Body Voltage Surges on Exceptional Subjects	W. Tiller et al.
	Six Modern Apparitional Experiences	I. Stevenson
	Viewing the Future: A Pilot Study with an Error-Detecting Protocol	R. Targ et al.
	Could Extraterrestrial Intelligences Be Expected to Breathe Our Air?	M. Swords
9:4	Decision Augmentation Theory: Applications to Random Number Generators	E. May
	Extrasensory Perception of Subatomic Particles & Referee Interchange (Dobyns)	S. Phillips
	North American Indian Effigy Mounds	A. Apostol
	A Holistic Aesthetic for Science	B. Kirchoff
10:1	An Assessment of the Evidence for Psychic Functioning	J. Utts
	Evaluation of a Program on Anomalous Mental Phenomena	R. Hyman
	CIA-Initiated Remote Viewing Program at Stanford Research Institute	H. Puthoff
	Remote Viewing at Stanford Research Institute in the 1970s: A Memoir	R. Targ
	American Institutes for Research Review of the STAR GATE Program	E. May
	FieldREG Anomalies in Group Situations	R. Nelson et al.
	Anomalous Organization of Random Events by Group Consciousness	D. Radin et al.
10:2	Critical Review of the "Cold Fusion" Effect	E. Storms
	Do Nuclear Reactions Take Place Under Chemical Stimulation?	J. Bockris et al.
	Claimed Transmutation of Elements Caused by a Chemical Process	V. Noninski et al.

	Selection versus Influence Revisited: New Methods and Conclusions	Y. Dobyns
	Illegitimate Science? A Personal Story	B. Maccabee
	Anomalous Phenomena Observed in the Presence of a Brazilian "Sensitive"	S. Krippner et al.
10:3	Mass Modification Experiment Definition Study	R. Forward
	Atmospheric Mass Loss on Mars and the Consequences	H. Lammer
	Exploring Correlations between Local Emotional and Global Emotional Events	D. Bierman
	Archetypes, Neurognosis and the Quantum Sea	C. Laughlin
10:4	Distance Healing of Patients with Major Depression	B. Greyson
	Cases of the Reincarnation Type: Evaluation of Some Indirect Evidence	J. Keil
	Enhanced Congruence between Dreams and Distant Target Material	S. Krippner et al.
	Recent Responses to Survival Research (Responses by Braude & Wheatley)	R. Almeder
	Toward a Philosophy of Science in Women's Health Research	A. Lettieri
11:1	Biased Data Selection in Mars Effect Research	S. Ertel/K. Irving
	Is the "Mars Effect" Genuine?	P. Kurtz et al.
	Fortean Phenomena on Film: Evidence or Artifact?	R. Lange/J. Houran
	Wishing for Good Weather: A Natural Experiment in Group Consciousness	R. Nelson
	Empirical Evidence for a Non-Classical Experimenter Effect	H. Walach/ S. Schmidt
	Consciousness, Causality, and Quantum Physics	D. Pratt
11:2	Anomalous Cognition Experiments and Local Sidereal Time	S. J. P. Spottiswoode
	Evidence that Objects on Mars are Artificial in Origin	M. Carlotto
	The Astrology of Time Twins: A Re-Analysis & Referee Interchange (Roberts)	C. French et al.
	Unconscious Perception of Future Emotions: An Experiment in Presentiment	D. Radin
	A Bayesian Maximum-Entropy Approach to Hypothesis Testing	P. Sturrock
	Planetary Diameters in the Surya-Siddhanta	R. Thompson
	Science of the Subjective	R. Jahn/B. Dunne
11:3	Accessing Anomalous States of Consciousness with Binaural Beat Technology	F. Holmes Atwater
	The "Mars Effect" As Seen by the Committee PARA	J. Dommanget
	Astrology and Sociability: A Comparative Psychological Analysis	S. Fuzeau-Braesch
	Comparison between Children with and without Previous-Life Memories	E. Haraldsson
	Did Life Originate in Space? Discussion of Implications of Recent Research	A. Mugan
	Correlations of Random Binary Sequences with Pre-Stated Operator Intention	R. Jahn et al.
	The Hidden Side of Wolfgange Pauli: An Encounter with Depth Psychology	Atmanspacher/ Primas
11:4	Topographic Brain Mapping of UFO Experiencers	N. Don/G. Moura
	Toward a Model Relating Empathy, Charisma, and Telepathy	J. Donovan
	The Zero-Point Field and the NASA Challenge of Create the Space Drive	B. Haisch/A. Rueda
	Motivation and Meaningful Coincidence: Further Examination of Synchronicity	T. Rowe et al.
	A Critique of Arguments Offered against Reincarnation	R. Almeder
	The Archaeology of Consciousness	P. Devereux
12:1	Gender Differences in Human/Machine Anomalies	B. Dunne
	Statement Validity Analysis of "Jim Ragsdale Story": Roswell Implications	J. Houran/S. Porter
	Experiment Effects in Scientific Research: How Widely Are They Neglected?	R. Sheldrake
	Roswell—Anatomy of a Myth	K. Jeffery
	A Different View of "Roswell—Anatomy of a Myth"	M. Swords
	Critique of "Roswell—Anatomy of a Myth"	R. Woods
12:2	Physical Evidence Related to UFO Reports	P. A. Sturrock et al.
	Empirical Evidence Against Decision Augmentation Theory	Y. Dobyns/R. Nelson
	Cases of Reincarnation in Northern India with Birthmarks and Birth Defects	S. Pasricha
	Can the Vacuum Be Engineered for Spaceflight Applications? Overview.	H. E. Puthoff
	Four Paradoxes Involving the Second Law of Thermodynamics	D. Sheehan
	The Paranormal Is Not Excluded from Physics	O. Costa de Beauregard

Index of Previous Articles in JSE

12:3 Estimates of Optical Power Output in Six Cases of Unexplained Aerial Objects J. Vallee
Analyses in Ten Cases of Unexplained Aerial Objects with Material Samples J. Vallee
Do Near-Death Experiences Provide Evidence for Survival of Human Personality E. Cook et al.
Anomalous Statistical Influence Depends on Details of Random Process M. Ibison
FieldREG II: Consciousness Field Effects: Replications and Explorations R. D. Nelson et al.
Biological Effects of Very Low Frequency (VLF) Atmospherics in Humans A. Schienle et al.
12:4 The Timing of Conscious Experience: Causality-Violating F. A. Wolf
Double-Slit Diffraction Experiment of Investigate Consciousness Anomalies M. Ibison/S. Jeffers
Techno-Dowsing: A Physiological Response System to Improve Psi Training P. Stevens
Physical Measurement of Episodes of Focused Group Energy W. Rowe
Experimental Studies of Telepathic Group Communication of Emotions J. Dalkvist/
  Westerlund

Strategies for Dissenting Scientists B. Martin
13:1 Significance Levels for the Assessment of Anomalous Phenomena R. A. J. Matthews
Retrotransposons as Engines of Human Bodily Transformation C. A. Kelleher
A Rescaled Range Analysis of Random Events F. Pallikari/E. Boller
Subtle Domain Connections to the Physical Domain Aspect of Reality W. A. Tiller
Parapsychology in Intelligence: A Personal Review and Conclusions K. A. Kress
Dreaming Consciousness: More Than a Bit Player in the Mind/Body Problem M. Ullman
13:2 The Effect of "Healing with Intent" on Pepsin Enzyme Activity T. Bunnell
Electronic Device-Mediated pH Changes in Water W. Dibble/W. Tiller
Variations on the Foundations of Dirac's Quantum Physics J. Edmonds
Do Cases of the Reincarnation Type Show Similar Features over Many Years? J. Keil/I. Stevenson
Optical Power Output of an Unidentified High Altitude Light Source B. Maccabee
Registration of Actual and Intended Eye Gaze: Correlation with Spiritual Beliefs G. Schwartz/
  L. Russek

Real Communication? Report on a SORRAT Letter-Writing Experiment I. Grattan-Guinness
What are the Irreducible Components of the Scientific Enterprise? I. Stevenson
Anomalies in the History of Relativity I. McCausland
Magic of Signs: A Nonlocal Interpretation of Homeopathy H. Walach
13:3 Second Sight and Family History: Pedigree and Segregation Analyses S. Cohn
Mound Configurations on the Martian Cydonia Plain H. Crater/
  S. McDaniel

Geomorphology of Selected Massifs on the Plains of Cydonia, Mars D. Pieri
Atmosphere or UFO? A Response to the 1997 SSE Review Panel Report B. Maccabee
An Unusual Case of Stigmatization M. Margnelli
Methuselah: Oldest Myth. or Oldest Man? L. McKague
Analysis of Technically Inventive Dream-Like Mental Imagery B. Towe/
  Randall-May

Exploring the Limits of Direct Mental Influence: Two Studies C. Watt et al.
13:4 Experimental Systems in Mind–Matter Research R. Morris
Basic Elements and Problems of Probability Theory H. Primas
The Significance of Statistics in Mind–Matter Research R. Utts
Introductory Remarks on Large Deviations Statistics Amann/
  Atmanspacher

p-adic Information Spaces. Small Probabilities and Anomalous Phenomena A. Khrennikov
Towards an Understanding of the Nature of Racial Prejudice Hoyle/
  Wickramasinghe

Clyde Tombaugh, Mars and UFOs M. Swords
14:1 Investigating Deviations from Dynamical Randomness with Scaling Indices Atmanspacher et al.
Valentich Disappearence: New Evidence and New Conclusion R. Haines/P.
  Norman

Protection of Mice from Tularemia with Ultra-Low Agitated Dilutions W. Jonas/D. Dillner

	The Correlation of the Gradient of Shannon Entropy and Anomalous Cognition	Spottiswoode/Faith
	Contributions to Variance in REG Experiments: ANOVA Models	R. Nelson et al.
	Publication Bias: The "File-Drawer" Problem in Scientific Inference	J. Scargle
	Remote Viewing in a Group Setting	R. Targ/J. Katra
14:2	Overview of Several Theoretical Models on PEAR Data	Y. Dobyns
	The Ordering of Random Events by Emotional Expression	R. Blasband
	Energy, Fitness and Information-Augmented EMFs in *Drosophila melanogaster*	M. Kohane/ W. Tiller
	A Dog That Seems To Know When His Owner Is Coming Home	R. Sheldrake/ P. Smart
	What Can Elementary Particles Tell Us about the World in Which We Live?	R. Bryan
	Modern Physics and Subtle Realms: Not Mutually Exclusive	R. Klauber
14:3	Plate Tectonics: A Paradigm Under Threat	D. Pratt
	The Effect of the "Laying On of Hands" on Transplanted Breast Cancer in Mice	Bengston/Krinsley
	Stability of Assessments of Paranormal Connections in Reincarnation Type Cases	I. Stevenson/J. Keil
	ArtREG: A Random Event Experiment Utilizing Picture-Preference Feedback	R. G. Jahn et al.
	Can Population Growth Rule Out Reincarnation?	D. Bishai
	The Mars Effect Is Genuine	S. Ertel/K. Irving
	Bulky Mars Effect Hard To Hide	S. Ertel
	What Has Science Come to?	H. Arp
14:4	Mind/Machine Interaction Consortium: PortREG Replication Experiments	Jahn/Mischo/ Vaitl et al.
	Unusual Play in Young Children Who Claim to Remember Previous Lives	I. Stevenson
	A Scale to Measure the Strength of Children's Claims of Previous Lives	J. B. Tucker
	Reanalysis of the 1965 Heflin UFO Photos	Druffel/Wood/ Kelson
	Should You Take Aspirin To Prevent Heart Attack?	J. M. Kauffman
15:1	The Biomedical Significance of Homocysteine	K. McCully
	20th and 21st Century Science: Reflections and Projections	R. G. Jahn
	To Be Or Not To Be! A 'Paraphysics' for the New Millennium	J. E. Beichler
	Science of the Future in Light of Alterations of Consciousness	I. Baruss
	Composition Analysis of the Brazil Magnesium	P. A. Sturrock
	Does Recurrent ISP Involve More Than Cognitive Neuroscience?	J.-C. Terrillon/ S. Marques Bonham
15:2	The Scole Investigation: Critical Analysis of Paranormal Physical Phenomena	M. Keen
	Bio-photons and Bio-communication	R. VanWijk
	Scalar Waves: Theory and Experiments	K. Meyl
	Commentary: On Existence of K. Meyl's Scalar Waves	G. W. Bruhn
	Cases of the Reincarnation Type in South India: Why So Few Reports?	S. K. Pasricha
	Mind, Matter, and Diversity of Stable Isotopes	J. P. Pui/A. A. Berezin
	Are the Apparitions of Medjugorge Real?	J. P. Pandarakalam
	Where Do We File 'Flying Saucers'? Archivist and Uncertainty Principle	H. Evans
	The Bakken: A Library and Museum of Electricity in Life	D. Stillings
15:3	A Modular Model of Mind/Matter Manifestations (M5)	R. G. Jahn/B. J. Dunne
	The Speed of Thought: Complex Space–Time Metric and Psychic Phenomenon	E. A. Rauscher/ R. Targ
	Failure to Replicate Electronic Voice Phenomenon	I. Baruss
	Experimental Study on Precognition	Vasilescu/Vasilescu
	Unexplained Temporal Coincidence of Crystallization	Constain/Davies
15:4	The Challenge of Consciousness	R. G. Jahn

Index of Previous Articles in JSE

	Anomalies and Surprises	H. H. Bauer
	Earth Geodynamic Hypotheses Updated	N. C. Smoot
	Unexplained Weight Gain Transients at the Moment of Death	L. E. Hollander, Jr.
	Physico-Chemical Properties of Water Following Exposure to Resonant Circuits	C. Cardella et al.
16:1	Can Physics Accommodate Clairvoyance, Precognition, and Psychokinesis?	R. Shoup
	The Pineal Gland and the Ancient Art of Iatromathematica	F. McGillion
	Confounds in Deciphering the Ramey Memo from the Roswell UFO Case	J. Houran/ K. D. Randle
	The Pathology of Organized Skepticism	L. D. Leiter
	Aspects of the Wave Mechanics of Two Particles in a Many Body Quantum System	Y. S. Jain
	Microscopic Theory of a System of Interacting Bosons: A Unifying New Approach	Y. S. Jain
	Unification of the Physics of Interacting Bosons and Fermions	Y. S. Jain
	The Pathology of Organized Skepticism	L. D. Leiter
16:2	Arguing for an Observational Theory of Paranormal Phenomena	J. M. Houtkooper
	Differential Event-Related Potentials to Targets and Decoys in Guessing Task	McDonough/Don/ Warren
	Stigmatic Phenomena: An Alleged Case in Brazil	S. Krippner
	The Case for the Loch Ness "Monster": The Scientific Evidence	H. H. Bauer
	What's an Editor To Do?	H. H. Bauer
16:3	M*: Vector Representation of the Subliminal Seed Regime of M5	R. G. Jahn
	Can Longitudinal Electromagnetic Waves Exist?	G. W. Bruhn
	Development of Certainty about the Deceased in Reincarnation Case in Lebanon	Haraldsson/ Izzeddin
	Manifestation and Effects of External Qi of Yan Xin Life Science Technology	Yan et al.
	Face-Like Feature at West Candor Chasma, Mars MGS Image AB 108403	Crater/Levasseur
	A Search for Anomalies	W. R. Corliss
	Common Knowledge about the Loch Ness Monster: Television, Videos, and Film	H. H. Bauer
16:4	Relationships Between Random Physical Events and Mass Human Attention	D. Radin
	Coherent Consciousness and Reduced Randomness: Correlations on 9/11/2001	R. D. Nelson
	Was There Evidence of Global Consciousness on September 11, 2001?	J. Scargle
	A Dog That Seems To Know When His Owner Is Coming Home	D. Radin
	An Investigation on the Activity Pattern of Alchemical Transmutations	J. Pérez-Pariente
	Anomalies in Relativistic Rotation	R. D. Klauber
	The Vardøgr, Perhaps Another Indicator of the Non-Locality of Consciousness	L. D. Leiter
	Review of the Perrott-Warrick Conference Held at Cambridge 3–5 April 2000	B. Carr
	Wavelike Coherence and CPT Invariance: Sesames of the Paranormal	O. Costa de Beauregard
	Why Only 4 Dimensions Will Not Explain Relationships in Precognition	Rauscher/Targ
17:1	Problems Reporting Anomalous Observations in Anthropology	C. Richards
	The Fringe of American Archaeology	A. B. Kehoe
	Rocks That Crackle and Sparkle and Glow: Strange Pre-Earthquake Phenomena	F. T. Freund
	Poltergeists, Electromagnetism and Consciousness	W. G. Roll
	AIDS: Scientific or Viral Catastrophe?	N. Hodgkinson
17:2	Information and Uncertainty in Remote Perception Research	B. J. Dunne/R. G. Jahn
	Problems of Reproducibility in Complex Mind–Matter Systems	H. Atmanspacher
	Parapsychology: Science or Pseudo-Science?	M.-C. Mousseau
	The Similarity of Features of Reincarnation Type Cases Over Many Years: A Third Study	I. Stevenson/ E. Haraldsson
	Communicating with the Dead: The Evidence Ignored. Why Paul Kurtz is Wrong	M. Keen
	Purported Anomalous Perception in a Highly Skilled Individual: Observations, Interpretations, Compassion	G. E. Schwartz/ L. A. Nelson/L. G Russek

	Proof Positive—Loch Ness Was an Ancient Arm of the Sea	F. M. Dougherty
17:3	Radiation Hormesis: Demonstrated, Deconstructed, Denied, Dismissed, and Some Implications for Public Policy	J. M. Kauffman
	Video Analysis of an Anomalous Image Filmed during Apollo 16	H. Nakamura
	The Missing Science of Ball Lightning	D. J. Turner
	Pattern Count Statistics for the Analysis of Time Series in Mind–Matter Studies	W. Ehm
	Replication Attempt: No Development of pH or Temperature Oscillations in Water Using Intention Imprinted Electronic Devices	L. I. Mason/ R. P. Patterson
	Three Cases of the Reincarnation Type in the Netherlands	T. Rivas
17:4	Testing a Language-Using Parrot for Telepathy	R. Sheldrake/A. Morgana
	Skin Conductance Prestimulus Response: Analyses, Artifacts and a Pilot Study	S. J. P. Spottiswode /E. C. May
	Effects of Frontal Lobe Lesions on Intentionality and Random Physical Phenomena Physical Phenomena	M. Freedman/S. Jeffers/K. Saeger/ /M. Binns/S. Black
	The Use of Music Therapy as a Clinical Intervention for Physiologist Functional Adaptation Media Coverage of Parapsychology and the Prevalence of Irrational Beliefs	D. S. Berger/ D. J. Schneck/ M.-C. Mousseau
	The Einstein Mystique	I. McCausland
18:1	A Retrospective on the *Journal of Scientific Exploration*	B. Haisch/M. Sims
	Anomalous Experience of a Family Physician	J. H. Armstrong, Sr.
	Historical Overview & Basic Facts Involved in the Sasquatch or Bigfoot Phenomenon	J. Green
	The Sasquatch: An Unwelcome and Premature Zoological Discovery?	J. A. Bindernagel
	Midfoot Flexibility, Fossil Footprints, and Sasquatch Steps: New Perspectives on the Evolution of Bipedalism	D. J. Meldrum
	Low-Carbohydrate Diets	J. M. Kauffman
18:2	Analysis of the Columbia Shuttle Disaster— Anatomy of a Flawed Investigation in a Pathological Organization	J. P. MacLean/ G. Campbell/ S. Seals
	Long-Term Scientific Survey of the Hessdalen Phenomenon	M. Teodorani
	Electrodermal Presentiments of Future Emotions	D. I. Radin
	Intelligent Design: Ready for Prime Time?	A. D. Gishlick
	On Events Possibly Related to the "Brazil Magnesium"	P. Kaufmann/ P. A. Sturrock
	Entropy and Subtle Interactions	G. Moddel
	"Can a Single Bubble Sink a Ship?"	D. Deming
18:3	The MegaREG Experiment	Y. H. Dobyns et al.
	Replication and Interpretation Time-Series Analysis of a Catalog of UFO Events: Evidence of a Local-Sidereal-Time Modulation	P. A. Sturrock
	Challenging Dominant Physics Paradigms	J. M. Campanario/ B. Martin
	Ball Lightning and Atmospheric Light Phenomena: A Common Origin?	T. Wessel-Berg
18:4	Sensors, Filters, and the Source of Reality	R. G. Jahn/ B. J. Dunne
	The Hum: An Anomalous Sound Heard Around the World	D. Deming
	Experimental Test of Possible Psychological Benefits of Past-Life Regression	K. Woods/I. Baruss
	Inferences from the Case of Ajendra Singh Chauhan: The Effect of Parental Questioning, of Meeting the "Previous Life" Family, an Attempt To Quantify Probabilities, and the Impact on His Life as a Young Adult	A. Mills
	Science in the 21st Century: Knowledge Monopolies and Research Cartels	H. H. Bauer
	Organized Skepticism Revisited	L. D. Leiter

19:1	The Effect of a Change in Pro Attitude on Paranormal Performance: A Pilot Study Using Naive and Sophisticated Skeptics	L. Storm/ M. A. Thalbourne
	The Paradox of Planetary Metals	Y. Almirantis
	An Integrated Alternative Conceptual Framework to Heat Engine Earth, Plate Tectonics, and Elastic Rebound	S. T. Tassos/ D. J. Ford
	Children Who Claim to Remember Previous Lives: Cases with Written Records Made before the Previous Personality Was Identified	H. H. Jürgen Keil/ J. B. Tucker
19:2	Balls of Light: The Questionable Science of Crop Circles	F. Grassi/C. Cocheo/ P. Russo
	Children of Myanmar Who Behave like Japanese Soldiers: A Possible Third Element in Personality	I. Stevenson/J. Keil
	Challenging the Paradigm	B. Maccabee
	The PEAR Proposition	R. G. Jahn/B. J. Dunne
	Global Warming, the Politicization of Science, and Michael Crichton's State of Fear	D. Deming
19:3	A State of Belief Is a State of Being	Charles Eisenstein
	Anomalous Orbic "Spirit" Photographs? A Conventional Optical Explanation	G. E. Schwartz/ K. Creath
	Some Bodily Malformations Attributed to Previous Lives	S. K. Pasricha et al.
	A State of Belief Is a State of Being	C. Eisenstein
	HIV, As Told by Its Discoverers	H. H. Bauer
	Kicking the Sacred Cow: Questioning the Unquestionable and Thinking the Impermissible	H. H. Bauer
19:4	Among the Anomalies	J. Clark
	What Biophoton Images of Plants Can Tell Us about Biofields and Healing	K. Creath/ G. E. Schwartz
	Demographic Characteristics of HIV: I. How Did HIV Spread?	H. H. Bauer
20:1	Half a Career with the Paranormal	I. Stevenson
	Pure Inference with Credibility Functions	M. Aickin
	Questioning Answers on the Hessdalen Phenomenon	M. Leone
	Hessdalen Research: A Few Non-Questioning Answers	M. Teodorani
	Demographic Characteristics of HIV: II. How Did HIV Spread	H. H. Bauer
	Organized Opposition to Plate Techtonics: The New Concepts in Global Tectonics Group	D. Pratt
20:2	Time-Normalized Yield: A Natrual Unit for Effect Size in Anomalies Experiments	R. D. Nelson
	The Relative Motion of the Earth and the Ether Detected	S. J. G. Gift
	A Unified Theory of Ball Lightning and Unexplained Atmospheric Lights	P. F. Coleman
	Experimenter Effects in Laboratory Tests of ESP and PK Using a Common Protocol	C. A. Roe/ R. Davey/P. Stevens
	Demographic Characteristics of HIV: III. Why Does HIV Discriminate by Race	H. H. Bauer
20:3	Assessing the Evidence for Mind–Matter Interaction Effects	D. Radin et al.
	Experiments Testing Models of Mind–Matter Interaction	D. Radin
	A Critique of the Parapsychological Random Number Generator Meta-Analyses of Radin and Nelson	M. H. Schub
	Comment on: "A Critique of the Parapsychological Random Number Generator Meta-Analyses of Radin and Nelson"	J. D. Scargle
	The Two-Edged Sword of Skepticism: Occam's Razor and Occam's Lobotomy	H. H. Bauer
20:4	Consciousness and the Anomalous Organization of Random Events: The Role of Absorption	L. A. Nelson/ G. E.Schwartz
	Ufology: What Have We Learned?	M. D. Swords
21:1	Linking String and Membrane Theory to Quantum Mechanics & Special	M. G. Hocking

	Relativity Equations, Avoiding Any Special Relativity Assumptions	
	Response of an REG-Driven Robot to Operator Intention	R. G. Jahn et al.
	Time-Series Power Spectrum Analysis of Performance in Free Response	P. A. Sturrock/
	Anomalous Cognition Experiments	S. J. Spottiswoode
	A Methodology for Studying Various Interpretations of the	M. A. Rodriguez
	N,N-dimethyltryptamine-Induced Alternate Reality	
	An Experimental Test of Instrumental Transcommunication	I. Barušs
	An Analysis of Contextual Variables and the Incidence of Photographic	D. B. Terhune et al.
	Anomalies at an Alleged Haunt and a Control Site	
	The Function of Book Reviews in Anomalistics	G. H. Hövelmann
	Ockham's Razor and Its Improper Use	D. Gernert
	Science: Past, Present, and Future	H. H. Bauer
21:2	The Role of Anomalies in Scientific Exploration	P. A. Sturrock
	The Yantra Experiment	Y. H. Dobyns et al.
	An Empirical Study of Some Astrological Factors in Relation to Dog Behaviour	S. Fuzeau-Braesch/
	Differences by Statistical Analysis & Compared with Human Characteristics	J.-B. Denis
	Exploratory Study: The Random Number Generator and Group Meditation	L. I. Mason et al.
	Statistical Consequences of Data Selection	Y. H. Dobyns
21:3	Dependence of Anomalous REG Performance on Run length	R. G. Jahn/
		Y. H. Dobyns
	Dependence of Anomalous REG Performance on Elemental Binary Probability	R. G. Jahn/
		J. C. Valentino
	Effect of Belief on Psi Performance in a Card Guessing Task	K. Walsh/
		G. Moddel
	An Automated Online Telepathy Test	R. Sheldrake/
		M. Lambert
	Three Logical Proofs: The Five-Dimensional Reality of Space–Time	J. E. Beichler
	Children Who Claim to Remember Previous Lives: Past, Present, & Future Research	J. B. Tucker
	Memory and Precognition	J. Taylor
	AIDS, Cancer and Arthritis: A New Perspective	N. Hodgkinson
	Online Historical Materials about Psychic Phenomena	C. S. Alvarado
21:4	Synthesis of Biologically Important Precursors on Titan Sam	H. Abbas/
	Is the Psychokinetic Effect as Found with Binary Random Number	D. Schulze-
	Generators Suitable to Account for Mind–Brain Interaction?	Makuch/
		Wolfgang Helfrich
	Explorations in Precognitive Dreaming	Dale E. Graff
	Climate Change Reexamined	Joel M. Kauffman
	Franklin Wolff's Mathematical Resolution of Existential Issues	Imants Barušs
	From Healing to Religiosity	Kevin W. Chen
22:1	Theme and Variations: The Life and Work of Ian Stevenson	Emily Williams
		Kelly/
		Carlos S. Alvarado
	Ian Stevenson: Recollections	Kerr L. White
	Reflections on the Life and Work of Ian Stevenson	Alan Gauld
	Ian Stevenson and Cases of the Reincarnation Type	Jim B. Tucker
	Ian Stevenson and the Modern Study of Spontaneous ESP Experiences	Carlos S. Alvarado/
		Nancy L. Zingrone
	Ian Stevenson's Contributions to Near-Death Studies	Bruce Greyson
	Ian Stevenson's Contributions to the Study of Mediumship	Erlendur
		Haraldsson
	Where Science and Religion Intersect: The Work of Ian Stevenson	Edward F. Kelly/
		Emily Williams

Index of Previous Articles in JSE 379

	Kelly
The Gentle American Doctor	M.M. Abu-Izzeddin
Professor Ian Stevenson—Some Personal Reminiscences	Mary Rose Barrington
Ian Stevenson: A Recollection and Tribute	Stephen E. Braude
Ian Stevenson and His Impact on Foreign Shores	Bernard Carr
Ian Stevenson: Gentleman and Scholar	Lisette Coly
The Quest for Acceptance	Stuart J. Edelstein
Ian Stevenson: Founder of the Scientific Investigation of Human Reincarnation	Doris Kuhlmann-Wilsdorf
Remembering My Teacher	L. David Leiter
Comments on Ian Stevenson, M.D., Director of the Division of Personality Studies and Pioneer of Reincarnation Research	Antonia Mills
Ian Stevenson: Reminiscences and Observations	John Palmer
Dr. Ian Stevenson: A Multifaceted Personality	Satwant K. Pasricha
A Good Question	Tom Shroder
The Fight for the Truth	John Smythies
Ian Stevenson: A Man from Whom We Should Learn	Rex Stanford
Ian Stevenson and the Society for Scientific Exploration	Peter A. Sturrock
Ian Stevenson's Early Years in Charlottesville	Ruth B. Weeks
Tribute to a Remarkable Scholar	Donald J. West
An Ian Stevenson Remembrance	Ray Westphal
22:2 Meditation on Consciousness	I. Ivtzan
An Exploration of Degree of Meditation Attainment in Relation to Psychic Awareness with Tibetan Buddhists	S. M. Roney-Dougal/ J. Solfvin/J. Fox
Thematic Analysis of Research Mediums' Experiences of Discarnate Communcation	A. J. Rock/J Beischel/ G. E. Schwartz
Change the Rules!	R. G. Jahn/ B. J. Dunne
Proposed Criteria for the Necessary Conditions for ShamanicJourneying Imagery	A. J. Rock/S. Krippner
"Scalar Wave Effects according to Tesla" & "Far Range Transponder"by K. Meyl	D. Kühlke
How to Reject Any Scientific Manuscript	D. Gernert
22:3 Unusual Atmospheric Phenomena Observed Near the Channel Islands, United Kingdom, 23 April 2007	J.-F. Baure/ D. Clarke/ P. Fuller/M. Shough
The GCP Event Experiment: Design, Analytical Methods, Results	P. Bancel/R. Nelson
New Insights into the Links between ESP and Geomagnetic Activity	Adrian Ryan
Phenomenology of N,N-Dimethyltryptamine Use: A Thematic Analysis	C. Cott/A. Rock
Altered Experience Mediates the Relationship between Schizotypy and Mood Disturbance during Shamanic-Like Journeying	A. Rock/G. Abbott/ N. Kambouropoulos
Persistence of Past-Life Memories: Study of Adults Who Claimed in Their Childhood To Remember a Past Life	E. Haraldsson
22:4 Energy, Entropy, and the Environment (How to Increase the First by Decreasing the Second to Save the Third)	D. P. Sheehan
Effects of Distant Intention on Water Crystal Formation: A Triple-Blind Replication	D. Radin/N. Lund/ M. Emoto/T. Kizu
Changes in Physical Strength During Nutritional Testing	C. F. Buhler/ P. R. Burgess/ E. VanWagoner
Investigating Scopesthesia: Attentional Transitions, Controls and	Rupert Sheldrake/

	Error Rates in Repeated Tests	Pamela Smart
	Shakespeare: The Authorship Question, A Bayesian Approach	P. A. Sturrock
	An Anomalous Legal Decision	R. A. Blasband
23:1	A New Experimental Approach to Weight Change Experiments at the Moment of Death with a Review of Lewis E. Hollander's Experiments on Sheep	Masayoshi Ishida
	An Automated Test for Telepathy in Connection with Emails	R. Sheldrake/ L. Avraamides
	Brain and Consciousness: The Ghost in the Machines	John Smythies
	In Defense of Intuition: Exploring the Physical Foundations of Spontaneous Apprehension	Ervin Laszlo
23:2	Appraisal of Shawn Carlson's Renowned Astrology Tests	Suitbert Ertel
	A Field-Theoretic View of Consciousness: Reply to Critics	D. W. Orne-Johnson/ Robert M. Oates
	Super-Psi and the Survivalist Interpretation of Mediumship	Michael Sudduth
	Perspectival Awareness and Postmortem Survival	Stephen E. Braude
23:3	Exploratory Evidence for Correlations between Entrained Mental Coherence and Random Physical Systems	Dean Radin/ F. Holmes Atwater
	Scientific Research between Orthodoxy and Anomaly	Harald Atmanspacher
23:4	Cold Fusion: Fact or Fantasy?	M. E. Little/S. R. Little
	"Extraordinary Evidence" Replication Effort	M. E. Little/S. R. Little
	Survey of the Observed Excess Energy and Emissions in Lattice-Assisted Nuclear Reactions	Mitchell R. Swartz
24:1	Rebuttal to Claimed Refutations of Duncan MacDougall's Experiment on Human Weight Change at the Moment of Death	Masayoshi Ishida
	Unexpected Behavior of Matter in Conjunction with Human Consciousness	Dong Shen
	Randomized Expectancy-Enhanced Placebo-Controlled Trial of the Impact of Quantum BioEnergetics and Mental Boundaries on Affect	Adam J. Rock Fiona E. Permezel
	A Case of the Reincarnation Type in Turkey Suggesting Strong Paranormal Information Involvements	Jürgen Keil
	Questions of the Reincarnation Type	Jürgen Keil
	How To Improve the Study and Documentation of Cases of the Reincarnation Type? A Reappraisal of the Case of Kemal Atasoy	Vitor Moura Visoni

JOURNAL OF SCIENTIFIC EXPLORATION

ORDER FORM FOR 2008 and 2009 PRINT ISSUES

JSE Volumes 1–21 are available FREE at scientificexploration.org
(JSE Vol. 22–24 are free to SSE members at http://journalofscientificexploration.org/index.php/jse/login)
JSE Volumes 22–23 (2008 and 2009) are available for $20/issue; use form below
Volume 24, Issues 1 & 2 are available at Amazon.com for $23.99 each.

Year	Volume	Issue	Price	×	Quantity	= Amount
			$20			

Postage and Handling per Issue Order
Choose class of mail

First Class/Airmail		Second Class/Media Mail	
USA	$4	USA	$2
N & S America	$5	Canada	$2.50
Europe	$6	All Other	$3
All Other	$7		

Subtotal
Postage
ENCLOSED

Name
Address
Phone/Fax

**Send to: SSE
P.O. Box 1190
Tiburon CA 94920 USA**

Fax (1) (415) 435-1654
EricksonEditorial@att.net
Phone (1) (415) 435-1604

☐ I am enclosing a check or money order

☐ Please charge my credit card as follows: ☐ VISA   ☐ MASTERCARD

Card Number   Expiration

Signature

☐ Send me information on the Society for Scientific Exploration

JOURNAL OF SCIENTIFIC EXPLORATION

GIFT ISSUES AND SUBSCRIPTIONS

Single Issue: A single copy of this issue can be purchased at Amazon.com
Price: $23.99

Subscription: To send a gift subscription, fill out the form below.
Price: $75 for an individual; $135 for a library/institution/business

Gift Recipient Name _____
Gift Recipient Address _____
_____

☐ I wish to remain anonymous to the gift recipient.
☐ I wish to give a gift subscription to a library chosen by SSE.
☐ I wish to give myself a subscription.
☐ I wish to join SSE as an Associate Member for $75/yr, and receive this quarterly journal (plus the EdgeScience magazine and The Explorer newsletter).

Your Name _____
Your Address _____
_____

Send this form to: *Journal of Scientific Exploration*
Society for Scientific Exploration
P. O. Box 1190
Tiburon CA 94920 USA

Fax (1) (415) 435-1654
EricksonEditorial@att.net
Phone (1) (415) 435-1604

For more information about the Journal *and the Society, go to*
http://www.scientificexploration.org

SOCIETY FOR SCIENTIFIC EXPLORATION

JOIN THE SOCIETY AS A MEMBER

The Society for Scientific Exploration has four member types:

Associate Member ($75/year): Anyone who supports the goals of the Society is welcome. No application material is required.

Student Member ($35/year): Send proof of enrollment in an accredited educational institution.

Full Member ($95/year): Full Members may vote in SSE elections, hold office in SSE, and present papers at annual meetings. Full Members are scientists or other scholars who have an established reputation in their field of study. Most Full Members have: a) doctoral degree or equivalent; b) appointment at a university, college, or other research institution; and c) a record of publication in the traditional scholarly literature. Application material required: 1) Your curriculum vitae; 2) Bibliography of your publications; 2) Supporting statement by a Full SSE member; or the name of Full SSE member we may contact for support of your application.

Emeritus Member ($75/year): Full Members who are now retired persons may apply for Emeritus Status. Please send birth year, retirement year, and institution or company retired from.

All SSE members receive: the quarterly *Journal of Scientific Exploration* (*JSE*), access to online SSE Member Directory, quarterly *EdgeScience* online magazine, *The Explorer* online newsletter, notices of meetings, reduced fees for SSE meetings, access to SSE online services, searchable *Journal* articles for the current year plus the two previous years (2008–2010) on the JSE site by member password.

Your Name _____

_____

Email _____
Phone _____ Fax _____
Payment of _____ enclosed, *or*
    Charge My   VISA ☐   Mastercard ☐
    Card Number _____ Expiration _____

    Send this form to:
        Society for Scientific Exploration
        P. O. Box 1190
        Tiburon CA 94920 USA

        Fax (1) (415) 435-1654
        EricksonEditorial@att.net
        Phone (1) (415) 435-1604

For more information about the Journal *and the Society, go to*
http://www.scientificexploration.org

JOURNAL OF SCIENTIFIC EXPLORATION
A Publication of the Society for Scientific Exploration

Instructions to Authors
(Revised March 2010)

All correspondence and submissions should be directed to:

JSE Managing Editor, P. O. Box 1190, Tiburon, CA 94920, USA, *email:* EricksonEditorial@att.net, *Telephone:* (1) 415/435-1604, *fax:* (1) 415/435-1654

Please submit all manuscripts at http://journalofscientificexploration.org/index.php/jse/login (please note that "www" is NOT used in this address). This website provides directions for author registration and online submission of manuscripts. Full author's instructions are posted on the Society for Scientific Exploration's website, at http://www.scientificexploration.org/jse/author_instr.php for submission of items for publication in the *Journal of Scientific Exploration*. Before you submit a paper, please familiarize yourself with the *Journal* by reading articles from a recent issue. (The Spring 2010 issue, JSE 24:1, can be ordered at Amazon.com. Back issues can be ordered from SSE Treasurer John Reed at joreed43@yahoo.com, or by using the order form in the back of this issue.) Electronic files of text, tables, and figures at resolution of a minimum of 300 dpi (TIF preferred) will be required for online submission. You will also need to attest to a statement online that the article has not been previously published or posted on the Internet and that it is not currently submitted elsewhere for publication.

AIMS AND SCOPE: The *Journal of Scientific Exploration* publishes material consistent with the Society's mission: to provide a professional forum for critical discussion of topics that are for various reasons ignored or studied inadequately within mainstream science, and to promote improved understanding of social and intellectual factors that limit the scope of scientific inquiry. Topics of interest cover a wide spectrum, ranging from apparent anomalies in well-established disciplines to paradoxical phenomena that seem to belong to no established discipline, as well as philosophical issues about the connections among disciplines. The *Journal* publishes research articles, review articles, essays, book reviews, and letters or commentaries pertaining to previously published material.

REFEREEING: Manuscripts will be sent to one or more referees at the discretion of the Editor-in-Chief. Reviewers are given the option of providing an anonymous report or a signed report. In established disciplines, concordance with accepted disciplinary paradigms is the chief guide in evaluating material for scholarly publication. On many of the matters of interest to the Society for Scientific Exploration, however, consensus does not prevail. Therefore the *Journal of Scientific Exploration* necessarily publishes claimed observations and proffered explanations that will seem more speculative or less plausible than those appearing in some mainstream disciplinary journals. Nevertheless, those observations and explanations must conform to rigorous standards of observational techniques and logical argument. If publication is deemed warranted but there remain points of disagreement between authors and referee(s), the reviewer(s) may be given the option of having their opinion(s) published along with the article, subject to the Editor-in-Chief's judgment as to length, wording, and the like. The publication of such critical reviews is intended to encourage debate and discussion of controversial issues, since such debate and discussion offer the only path toward eventual resolution and consensus.

LETTERS TO THE EDITOR intended for publication should be clearly identified as such. They should be directed strictly to the point at issue, as concisely as possible, and will be published, possibly in edited form, at the discretion of the Editor-in-Chief.

PROOFS AND FINAL PDFs: After reviewing their copyedited manuscript, corresponding authors will receive typeset PDF page proofs for review. PDFs of the final published online version, and print Journals, are available to all named authors of an article upon request.

COPYRIGHT: Authors retain copyright to their writings. However, when an article has been submitted to the *Journal of Scientific Exploration* for consideration, the *Journal* holds first serial (periodical) publication rights. Additionally, after acceptance and publication, the Society has the right to post the article on the Internet and to make it available via electronic as well as print subscription. The material must not appear anywhere else (including on an Internet website) until it has been published by the *Journal* (or rejected for publication). After publication in the *Journal*, authors may use the material as they wish but should make appropriate reference to the prior publication in the *Journal*. For example: "Reprinted from [or From] "[title of article]", *Journal of Scientific Exploration*, vol. [xx], no. [xx], pp. [xx], published by the Society for Scientific Exploration, http://www.scientificexploration.org."

DISCLAIMER: While every effort is made by the Publisher, Editors, and Editorial Board to see that no inaccurate or misleading data, opinion, or statement appears in this *Journal*, they wish to point out that the data and opinions appearing in the articles and advertisements herein are the sole responsibility of the contributor or advertiser concerned. The Publisher, Editors, Editorial Board, and their respective employees, officers, and agents accept no responsibility or liability for the consequences of any such inaccurate or misleading data, opinion, or statement.